# FROM THE BIG BANG TO GOD

## Our Awe-Inspiring Journey of Evolution

## Lloyd Geering

**POLEBRIDGE PRESS**
Salem, Oregon

D0291443

Cover and interior design by Robaire Ream

**Library of Congress Cataloging-in-Publication Data**

Geering, Lloyd, 1918-
 From the big bang to God : our awe-inspiring journey of evolution / Lloyd Geering.
     pages cm
  Includes bibliographical references and index.
  ISBN 978-1-59815-139-8 (alk. paper)
1. Evolution--Religious aspects--Christianity. 2. Religion--Philosophy. I. Title.
  BL263.G44 2013
  200--dc23

                                                        2013023622

*To my great-grandchildren*
*Fynn and Neave, Maxim and Linus, Xavier and Anouk,*
*Amelia, Angus, and any yet to be born*

# Stages in Evolution

| | | |
|---|---|---|
| **Big Bang**<br>13.75 billion years ago | | **Big Bang**: emergence of the cosmos<br>From 13.75 billion years ago |
| **Hadean eon**<br>4.5 billion to 4 billion years ago | | **Geogenesis**: emergence of Earth<br>From 4.5 billion years ago |
| **Archean eon**<br>4 billion to 2.5 billion years ago | | **Biogenesis**: emergence of life<br><br>First cells<br>From 3 billion years ago |
| **Proterozoic eon**<br>2.5 billion to 541 million years ago | | More complex cells<br><br>First multi-cellular organisms |
| **Phanerozoic eon**<br>541 million years ago to present | **Paleozoic era**<br>Age of the Fishes<br>541 million to 252 million years ago | **Biodiversity of life**<br><br>Plant life invades the land<br>Insect life invades the land<br>Sea animals invade the land |
| | **Mesozoic era**<br>Age of the Reptiles<br>252 million to 66 million years ago | Life diversifies in sea and on land<br>Some reptiles take to the air as birds<br><br>The dinosaurs come to dominate<br>The earliest mammals appear |
| | **Cenozoic era**<br>Age of the Mammals<br>66 million years ago to present | **Anthropogenesis**: emergence of humankind<br>(From mammal to Homo sapiens)<br><br>**Logogenesis**: emergence of language<br>2,000,000 years ago<br>**Noogenesis**: emergence of the noosphere<br>50,000 years ago<br>**Polytheogenesis**: emergence of the gods<br>20,000 years ago<br>**Monotheogenesis**: emergence of God<br>2500 years ago<br>**Homogenesis**: 'God' comes down to earth<br>250 years ago |

# Contents

# Preface

*About two years ago*

I was invited to participate in a popular discussion forum on scientific subjects held monthly in the coffee lounge at Te Papa, the National Museum of New Zealand. The invitees gave a short introduction to the chosen topic before it was opened up to general discussion. The topic allotted to me concerned the relationship between science and religion.

I have been equally interested in science and religion throughout most of my working life, and almost opted for a career in science before being diverted into theology. I continued to read popular books on science, for I believe that theology must never be allowed to remain in conflict with the sciences, as it has all too often appeared to be since the mid-19th century.

In the early 1960s I read a recently published book on evolution by the French scientist and priest Pierre Teilhard de Chardin and found it so absorbing that I was hardly able to put it down. To appreciate evolution, we need to experience a sense of the passing of time over long periods, for a human life occupies such a short segment of cosmic time that it is impossible to observe biological evolution taking place. Teilhard took me through the various stages of the evolutionary process at such a breakneck speed that I found it a thrilling journey.

Remembering this experience, and believing that Teilhard had blended science and religion into one seamless story, I decided to take up the challenge of telling the story of evolution simply and concisely. But to do so in the fifteen minutes allowed by Te Papa's schedule seemed an impossible task. I had to race through the chief stages in the evolutionary process at an average speed of a billion years a minute to allow a necessary deceleration towards the end. It left my audience gasping for breath. That, of course, is how we *should* feel as we hear the story of evolution being unfolded.

The positive response from the audience encouraged me to extend my fifteen-minute sketch into this book. Some very fine books have already told the story of cosmic and biological evolution, but few if any link it with the evolution of human culture in a way that shows how they constitute one continuous story in which religion and science play their respective roles and are closely related to each other.

To tell that story required a great deal of research in areas where I had only general and often hazy knowledge—especially cosmology, geology, biology and linguistics. I have learned a great deal in the course of writing this book and I hope readers may find it interesting also. I believe and hope that by understanding and accepting what we now know of the evolutionary process, the human race will be able to attain a viable future for itself. I am fully aware that in covering a topic so vast there may be errors of detail, but I trust that any such infelicities will not obscure or invalidate the general stream of thought and argument.

Several people have assisted me: Hamish Campbell, the geologist who invited me to speak at Te Papa and thus unintentionally initiated this venture; David Burton, biologist; Michael Corballis, psychologist and linguist; and my theological friend David Simmers. All have kindly read the chapters relating to their areas of expertise and made valuable suggestions, and I thank them. Last but not least I am indebted to my indefatigable editor and master wordsmith, Tom Hall.

—*Lloyd Geering*

# Introduction

## Where Did We Come from?

*Among the most important questions
we can ask are:*

Who are we? Where did we come from? Why are we here? Where
are we going? And although many a curious child has asked,
"Where did I come from?", most of us reach adolescence before
we start pondering these questions in any depth. By that time,
we probably already absorbed some of the answers that our
common culture has produced and passed down over countless
generations.

Every culture has its own set of answers to these questions
and hands them on to each new generation. Such answers usually
explain the origin of a particular tribe or nation and sometimes
include stories of the origin of the human race as a whole. Thus,
each culture comes to possess its own version of what I refer to
as the *Great Story.*

Western culture, shaped for centuries by the Christian tradition, long assumed that the Bible supplied authoritative answers
to these questions. The Great Story proclaimed by Christians
declared that some six thousand years ago God made the world
and all that it includes—and did so in the short space of six days.
Only on the sixth day did God create two human beings, Adam
and Eve, from dust of the ground. For two thousand years few
Christians so much as questioned the notion that these were our
first parents and that all humans of every race and colour had
descended from them.

1

Only in the mid-nineteenth century did this Great Story find itself the object of sceptical analysis that led to it being gradually replaced by a different narrative, commonly referred to as the story of evolution. To be sure, conservative Christians reject the new story—either in its entirety or in part—and defend the traditional Great Story on the grounds that the Bible is the literal word of God. Thus, since the mid-nineteenth century, Western culture has found itself divided between the old and the new Great Story—for which reason, among others, our culture has lost much of the cohesion it once possessed.

Now in the twenty-first century, we find ourselves encountering problems that can no longer be solved by individual nations and cultures, but only by the whole human race working together: global warming, the pollution of air and water, and the population explosion that is already resulting in mass starvation. If the human race is to learn to work together, it must first acknowledge and understand its common origin. This is precisely what the new Great Story provides for us. Though most people are vaguely aware of this new way of looking at things and familiar with some aspects of it, the details of the story make it so long and complex that it usually remains on the periphery of our consciousness. Most of us spend so much of our time caught up with the details of our own little circles of interest that we rarely turn our attention to the ultimate questions of life. We are too busy living life to stop and ask who we are and why we are here.

To exemplify this, take the case of an endearing character from what is often claimed to be the most widely read book of the nineteenth century, Harriet Beecher Stowe's 1852 anti-slavery novel, *Uncle Tom's Cabin*. Topsy was a little slave girl who, on being asked if she had ever heard about God, looked bewildered. Her little white friend attempted to prompt her by saying, "Do you know who made you?" To this she replied, "Nobody, as I knows on, I s'pect I grow'd. Don't think nobody never made me." This book was so widely read that before long "it growed like Topsy" became a popular figure of speech that described anything that appeared to develop by itself without any apparent cause or designer.

By a strange coincidence this quaint phrase, here put into the mouth of a naïve and uneducated child, actually expresses with remarkable simplicity the essence of the epic tale of evolution, which tells how, beginning with the lifeless material (dust) of Earth, we humans simply 'grew' from simpler forms of life.

*Uncle Tom's Cabin* is said to have been a factor in what led to the civil war that was fought to abolish slavery in the United States. In 1859 Charles Darwin published *On the Origin of Species*, the book that led to the still hard-fought theological war between conservative and liberal Christians. Conservative Christians argue, perhaps understandably, that the idea of evolution undermines the validity of Christianity by rendering the idea of a Creator God unnecessary.

The question posed by Topsy's little friend neatly illustrates why those who stoutly defend the role of God as the divine Creator so flatly reject the story of evolution. To them it seems self-evident that everything that exists must have had a maker. They are convinced that not only the Bible but reason itself is on their side; as they see it, only the unsophisticated or irrational mind could accept the notion that the human race, and the universe itself, "just growed like Topsy".

Yet as the Bible puts it, "out of the mouth of babes and sucklings" may come the most amazing insights. This is not because the young and inexperienced possess some special source of knowledge, but rather because they may be open to ideas that older people are inclined to reject as impossible, as their imaginations are constrained by what is generally accepted as true in the real world.

All that Topsy knew about was 'growth', for had she not observed it taking place all around her? We place a tiny seed in the soil and a plant emerges. It may even grow into a huge tree that greatly outlives *us*. We break a bone in our body and wait patiently for the two pieces to knit without pondering the mystery of how and why bones can *grow together* again. We observe with delight the way an infant progresses by a natural process of *growth* through childhood, adolescence and adulthood, without stopping to recognise what an extraordinary process it is.

To begin to understand the story of evolution, we need only ponder the mysterious phenomenon of growth, for it demonstrates before our very eyes the very essence of evolution. But because growth is all around us, and personally experienced as we attain adulthood, we pretty much take it for granted and therefore do not pause to wonder at the mystery of this phenomenon. We sometimes declare an unexpected deliverance from an incurable disease to be a miracle, but fail to recognise and acknowledge how much more miraculous is the everyday phenomenon of growth that we see in life all around us.

You and I were once newborn infants, completely helpless and knowing nothing. Before that, each of us was a fetus in our mother's womb, having grown from a very tiny egg, fertilised by an even tinier sperm donated by our father. What a miracle it is that we thinking, feeling, active humans grew from such a beginning. Indeed, what an awe-inspiring fact it is that we exist at all.

Even this concisely expressed description of where we came from is much more than Topsy possessed. Further, it is more than even most educated persons knew prior to the twentieth century. Therefore until quite recently people of every culture considered the origin of each person to be such a mystery that all kinds of stories were created and passed on from generation to generation to explain where we came from. A common one that lasted until fairly recent times was that God, the creator of all things, implanted the soul (the essence of a person) into the embryo at a certain time during pregnancy, an event referred to as the *quickening* (which simply means the 'enlivening') of the fetus.

During the twentieth century our whole understanding of the universe changed out of all recognition; our knowledge expanded a million-fold, even as to situations and happenings that could not be seen by the human eye, and for the most part never will be. Until the twentieth century we humans were aware only of what we may refer to as the *viewable world*. The human eye can see only those objects that fall within a particular range of size. The larger limit of this range includes the planets and brighter stars, though even they can be seen only because their enormous distance from us has reduced their apparent size to within the viewable range. At the smaller limit of the range we can see with the naked eye nothing beyond tiny insects, little seeds and specks of dust. Our ancestors, until quite recent times, understandably took this viewable world to be the whole universe. It now appears that the universe includes a great deal more than we can personally observe. And since we humans are quite massive creatures when compared with bacteria, I shall hereafter speak of the viewable world as the *macro-world*.

During the twentieth century scientists made us aware that beyond the limits of the macro-world exists both what we may call the *mega-world* and the *micro-world*. The mega-world consists of the space-time continuum now studied by cosmologists. It is so vast and has existed for so long that we have great difficulty in creating an image of it in our minds. To be sure, we see a tiny fragment of it as we view the night sky, but we depend on cosmologists to explain just what it is we are observing when we look at the sparkling stars—many of which, it turns out, are masses of stars called galaxies! This mega-world contains the macro-world that we are so familiar with, but it is 'astronomically' greater in size.

Another group of scientists (mainly physicists and biologists) have introduced us to the micro-world, the elements of which are so unbelievably small by comparison with the macro-world that once again we humans may never be able to see them with our own eyes. It is in this micro-world that we find the building blocks of the physical world and of living organisms—including the many kinds of viruses and bacteria that can cause suffering and death.

At the very lowest level of the micro-world is the *subatomic world*. From Democritus (c. 460–370 BCE) we have inherited the word *atom*, by which he referred to the smallest possible bit of physical material, one that cannot be broken down any further. The successful splitting of the atom by Ernest Rutherford (1871–1937) opened up to us the subatomic world of electrons, protons, neutrons and quarks that constitute the energy from which all physical material is composed.

Until recently our experience of the way stable matter operates was adequately explained by what we know as the laws of Newtonian physics, but we now find that energy at the subatomic level operates in ways that are very puzzling and even hard to believe. At this quantum level of reality, for example, some particles (tiny bits of energy) seem to operate more by chance than by any fixed law of nature. Since they cannot be seen, we have to create mathematical models to describe what takes place at the subatomic level. More perplexing, because subatomic matter sometimes manifests itself as wave motions and sometimes as minute particles, no single model or theory yet fully explains its behaviour.

The mega-world, the macro-world and the micro-world are one continuous reality that constitutes the universe. Of course, the personal knowledge and experience of laypeople like myself is restricted to the limits of the macro-world. Because we shall never see the micro-world with our own eyes nor view the universe as a whole, our knowledge of these worlds is wholly dependent on the expertise of scientists. They have replaced the priests and prophets on whom we assumed ourselves to be dependent in former times when much of our knowledge was thought to come by way of divine revelation.

Our present mutual dependence applies even to the experts themselves with regard to fields of study other than their own. So great has been the explosion of available knowledge in recent times, and so diversified has the scientific development become into new and previously unknown areas, that it is impossible for anyone to become an expert in more than one confined area. For reliable answers to the basic questions of human existence, humans of all races and cultures have become mutually dependent to a degree never before known.

The experts we depend on do not come from any one ethnic group or culture, but transcend those divisions that were so dominant in the past. Even though the scientific method of research emerged primarily in Western culture, it has left its cultural origins behind and is engaging the attention and participation of all ethnic groups. It is global in its outreach, and the rapid advance in human communication enables science to speed ahead at an accelerating rate. Its rapidly accumulating body of knowledge (now mostly accessible to everybody through the Internet) is providing material for the new Great Story that has the potential to unite the whole human race as it faces the challenges of the twenty-first century.

This book is an attempt to sketch that new Great Story of where we came from, who we are and where we are heading.

Let us start from where we now stand. Naturally enough, we are inclined to trace our personal beginnings to the date of our birth, an event we celebrate each year. But we already lived some nine months in our mother's womb following the moment of our conception. Further, the various physical features and traits that we inherited through our father's sperm and our mother's egg mean that the roots of our being go very much further back. Thus, our personal lifetime covers only a very short period when compared with the evolution of the genes that provide us with much of our unique identity.

The evolution of the human species so far is, in turn, only an infinitesimal segment in the unimaginably long history of life on this planet. So with regard to the question of where we came from, the explosion of knowledge in modern times has rendered completely obsolete the answers that past cultures created and handed on. Taken as a whole, science has transcended and in effect cancelled the racial and cultural differences of the past.

The Judaeo-Christian story of origins, with which most Westerners are familiar, declared that humans were made from the dust of Earth and that, as our funeral services sometimes make explicit, we eventually return to dust. In a strange and unexpected way, the new story of where we came from largely confirms this, for it tells how all life forms on this planet evolve out of the physical substance of Earth and return to it upon death. We humans are earthly creatures who have evolved out of Earth, and we share a common ancestry with all other forms of earthly life. We have evolved to fit the conditions of Earth, such as its particular gravitational pull, its atmosphere and its available water supply. Not only are we part of Earth, but Earth, in turn, is no more than the tiniest fragment of the vast and ever-evolving universe.

Pierre Teilhard de Chardin (1881–1955) was a scientist and priest who was equally committed both to science and to the Christian tradition in which he had been reared. He spent his life attempting to unite this new story of our origins with the spirituality of the old one. Because his thoughts were so blatantly in conflict with the teachings of his church, he was prevented from publishing them during his lifetime; only after his death (about a hundred years after *Uncle Tom's Cabin* and *On the Origin of Species*) did his magnum opus *The Phenomenon of Man* first see the light of day. Although widely read and admired at the time, it was also bitterly criticised by both scientists and traditional

Christians. True, it could not be accepted as the 'scientific treatise' Teilhard de Chardin thought it to be, but it drew on the scientific data of the day as no theological discourse had ever done before.

Teilhard's book is now somewhat dated, and its value must be assessed within its cultural context, for the biological sciences had not achieved today's scope and sophistication, and theologians were still struggling to come to terms with the implications of biological evolution. His book may best be understood as an awe-inspiring vision of how the ever-evolving universe unfolded itself, eventually bringing forth on planet Earth the self-aware animal species we know as humankind. It is the story both of how *we* came to be and of how, as it were, the evolutionary process is at last becoming conscious of itself. This latter point was made by Sir Julian Huxley, the eminent scientist who introduced Teilhard's book to the world with considerable approval.

Teilhard very rarely used the word *God* in his book; there was no need to, since for him the whole evolutionary process from beginning to end (or as he put it, "from alpha to omega") was the new way in which he was coming to understand the meaning of the word. Whereas tradition had spoken of God as the *Creator*, Teilhard focused the reader's attention on the ongoing *creativity* that permeates the universe, a phenomenon that continually unfolds itself in ever-more complex forms.

I refer to Teilhard because he coined or borrowed a number of terms that are still useful in helping us non-scientists understand the universe we live in. One of these is *cosmogenesis*, a word coined by Helena Blavatsky in 1888, which Teilhard used to refer to the nature of the universe as one continuous process of formation. The cosmos is not a static reality, as the ancients supposed, and as many people today tend to think it to be; it is a process of continuous change on the grand scale. It did not suddenly appear as a finished product. The creation of most things (including those we manufacture) is not a single event but a process of coming into being. To use Topsy's words, the universe was not "made", but rather "grow'd"; it has been evolving from a simple but most extraordinary beginning. And it is still evolving and growing as its expanding and ever-changing nature so clearly illustrates. That is why most of the chapter headings in this book include the root *-genesis*, a word intended to indicate process—the process of emergence or coming into being.

So to begin the Great Story of where we came from, we turn first to *cosmogenesis*—the emergence of the cosmos.

# The Evolution of
# the Physical Universe

# Cosmogenesis

## The Coming into Being
of the Universe

*From 13,750,000,000 years ago*

Until less than two hundred years ago, most people in the Western world firmly believed, on biblical authority, that Earth and sky were no more than six thousand years old. Then a dramatic change began to take place. First came the geologists: they began to show us, on the basis of the scientific study of rocks, that so far as Earth is concerned we must think in terms of at least hundreds of thousands of years. Hard on the heels of the geologists came the astronomers and cosmologists, who told us that the universe itself has been in existence for billions of years, and today they calculate it to be 13.75 billion years old—almost three times as old as Earth on which we live.

But can our minds really cope with such enormous periods of time? In a normal lifetime we humans experience only an infinitesimal fragment of cosmic time. Compared with such a relatively brief span of time, there seems little practical difference between a million years and a billion years—or for that matter, eternity itself. Indeed, relative to a person's lifetime, the period we can best understand, the difference between a billion (1,000,000,000) years and eternity seems only academic. That is why in this book I sometimes use figures rather than words—the number of zeros helps to bring home the length of the time periods we are dealing with.

At first cosmologists were led to consider the likelihood that the universe had always been in existence. They spoke of the cosmos as a steady-state universe in which heavenly bodies continuously appear and disappear, leaving the vast expanse and everything in it to have neither a beginning nor an end.

But in 1929 cosmologist Edwin Hubble found evidence that the stars, and indeed whole galaxies, are moving away from one another. What is more, the further away any observable galaxy is judged to be, the faster it is moving away from us. Thus cosmologists concluded that the universe is expanding. But the galaxies are not moving away from one another in an already-existing space. As we shall soon note, there is no such thing as absolute space. Space is simply the distance between objects, and the distance between objects in the sky is increasing. That is what is meant by saying that the universe is expanding. Moreover, space and time are intimately related to one another in such a way that we must now think of time as the fourth dimension of the universe. When we look out into space with our giant telescopes, we are also looking back in time; depending on what star or galaxy we are looking at, what we see is what used to be there thousands, millions, or even billions of years ago.

The reason why space can expand is that it behaves as though it is elastic; that is, three-dimensional space somewhat resembles the two-dimensional surface of a balloon that is being inflated. The way that heavenly bodies are moving away from one another in the expanding universe may be likened to the way the dots on the surface of an expanding balloon become further and further apart. This helps us to understand why cosmologists speak of the curvature of space. Today everyone knows that we can sail across the oceans in a straight line without fearing—as many of Columbus' sailors did only five hundred years ago—that we shall fall over the edge. For much the same reason, it is impossible to reach the edge of the universe: it has no edges. If we travel round our earthly globe in an apparently straight line we come back to where we started. So it is with the space-time universe.

Similarly, the universe has no centre, just as the surface of a balloon has no centre. Philosophers, theologians and scientists alike had long assumed that Earth occupied this pre-eminent location, but from Copernicus and Galileo onwards, we were led to think that the sun was the focal point of the solar system and the galaxy. For a very short time it even appeared that our galaxy marked the centre of all creation, but we can now say categorically that the universe has no centre. The many trillions of stars and galaxies that constitute the universe cannot be said to have a centre, for no matter what point the universe is viewed from, the rest of it appears to be more or less uniform and rapidly receding.

From Einstein we learned that time and space are interdependent. The *absolute space* that Isaac Newton assumed, and we find natural to imagine, simply does not exist. Neither is there any such thing as *absolute time*. Time and space are both relative to the viewer. That is why cosmologists now speak of the universe as the *space-time continuum*. Therefore, we cannot postulate an absolute present moment in the universe, for what we regard as the present is always relative to us and to where we are. This varies even for people on Earth, and may be illustrated by the short time lag experienced during international phone calls.

But how much more is this the case when we survey the starry heavens? As our telescopes enable us to look out into space, they are also taking us back into time. So great are the distances involved that we now measure them in the time it takes light to travel across them. The common unit used in cosmology is a *light year*, the distance that light—speeding at 186,000 miles per second—travels in the course of an earthly year.

But if the universe is expanding in the way cosmologists have found, we must theoretically be able to go back in time to a period when it was much smaller than it is today. Yet how far back can we go? Into how small a space could the universe be compressed? We come to a stage where we find ourselves dealing with processes and phenomena that no longer obey the known laws of Newtonian physics, the physics of our macro-world.

Because we cannot go back in time to view the beginning of the universe, and because our macro-world perspective prevents us from viewing the universe as a whole, scientists construct models and mathematical formulae to calculate what would happen if we reversed the process of expansion. These they then put to the test in order to judge their validity. If a model leads them to make accurate predictions, their confidence in it is confirmed. It is by such means that cosmologists have learned much about the universe of long ago, and the present consensus is that it had a definite beginning some 13.75 billion years ago. They speak of that beginning as the *Big Bang*.

This term originally expressed a somewhat derogatory dismissal by 'steady-state' theorists of the then-novel theory that the universe had a definite beginning. But the term stuck and is now widely used to describe the presumed 'event' from which the universe emerged. This 'event' is significantly different from any event that has since occurred, and for that reason it is referred to as a *singularity*. It is beyond our ability to imagine, for it has strange and seemingly illogical implications. If the Big Bang marked the beginning of the space-time continuum, then there was neither space nor time before this event: it marked the beginning of both time and space simultaneously. The fact that there was no time before the Big Bang makes it illogical to ask what led up to that event. So we cannot attribute the Big Bang to a First Cause (as Aristotle suggested) or to the workings of a divine will (as Christian tradition had long affirmed). Later we shall discuss how it was that 'God' came into the story, but for now it is sufficient to say that the nature of the Big Bang rules out any literal understanding of the biblical statement, "In the beginning God created the sky and the Earth".

So why did the primal event we now call the Big Bang ever take place? Philosophers have long discussed the question of the creation of the universe as the issue of why there is something and not nothing. The very character of the Big Bang as a singularity implies that this is a question that has no answer, that there *is* no reason or cause for the universe to exist. But if the Big Bang occurred for no reason, then the alternative must be given serious consideration: it was by sheer chance that the universe came into being. This possibility is not entirely absurd, for later in our story we shall find that chance plays a considerable role in the way the universe has evolved. All that we can say with confidence is, "The universe simply is"—in much the same way as each of us can say, "I am", knowing full well that we might very well never have been born.

Thus at the very beginning of time, while the universe was infinitesimally small, nothing was certain; rather, there suddenly came into existence an almost infinite array of possibilities. An infinitely powerful point of energy was pure potential. Only as some of the immense range of possibilities came to be realised did it increasingly take form and shape, including what we now call the very laws of nature by which the universe operates. Physicists have now concluded that the basic forces operating in the universe are no more than four: gravity, electromagnetism, the strong nuclear force and the weak nuclear force.

It is worth noting that decades earlier Teilhard de Chardin's gifted imagination had already led him to include a similar suggestion in *The Phenomenon of Man*. His observations, based on then-current knowledge of biological evolution, inclined him to propose the existence of two opposing cosmic forces that he termed *tangential energy* and *radial energy*. Teilhard chose the word *tangential* from the way an object loosely attached to a spinning wheel may fly off at a tangent. Tangential energy has an explosive (or expansive) effect and is exemplified in a nuclear bomb. Radial energy, on the other hand, has the effect of pulling things together. Its most common example is the phenomenon of gravitation, first isolated by Isaac Newton to explain why all objects with mass attract one another. (This notion has since been explained by the effect of the curvature of space posited by Einstein's theory of relativity.) Planets revolving around the sun stay in their orbits because the tangential energy created by their momentum is exactly balanced by the radial energy exerted by gravity.

Teilhard's suggested model of two opposing forms of energy must now be judged as simplistic and unscientific. Nevertheless, it remains useful in helping us non-scientists understand the basic principles of the evolving universe that have appeared again and again and in myriad ways from the moment of the Big Bang until now. To avoid confusion with the way scientists now speak of energy, it may be better to speak of the self-creative universe as showing two opposing tendencies—the spreading or explosive tendency and the gathering or uniting tendency. Examples of these opposing tendencies can be found in the way magnetic forces both attract and repel, in the opposition of positive and negative electric charges, and in the opposition of matter and antimatter (a topic we shall take up shortly).

The creativity that continually permeates the universe may be traced back to the interplay between these two opposing forces. Eventually they brought forth and sustained life; they may be seen in the way our lungs take in oxygen by alternately inflating and deflating, and in the way our hearts pump blood throughout our bodies by alternately expanding and contracting. The operation of these two opposing tendencies enables the universe to become a self-creative process that we can almost envision as an astronomically immense living entity.

But what actually happened at the beginning of this long creative process? While the term 'Big Bang' suggests an explosion, it was not an explosion of some primal combustible materials already existing in otherwise empty space. There is still much speculation about the state of the universe in its first three seconds. The universe seemingly originated as an unimaginably large bundle of energy so densely packed that it took up virtually no space at all. This means that it was subject to heat and pressure far beyond our gauges of measurement and even our imagination.

During the first few micro-seconds the universe expanded exponentially, and hence the appropriateness of the term 'Big Bang'. But then the expansion began to slow down, and as it did so, the universe began to cool. Thus, two opposing tendencies were in operation from the beginning. Even more surprisingly, scientists have found that if the rate of expansion had been any greater, the stellar clouds would not have formed, while if it had been even a tiny bit slower it would have collapsed upon itself and perhaps disappeared into nothing. The universe has been expanding at just the right rate to become what it is and to bring forth life. That in itself is a mystery worth pondering.

So what happened as it began to cool? At that primal stage the universe consisted simply of minute elements of energy that were moving and colliding with one another at such enormous speeds that creation and destruction were going on simultaneously. Out of this confusing turmoil there began to emerge what are now referred to as *matter* and *antimatter*. Thus, it now appears that the ancient biblical writer was not far from the right track when he described how an ordered world emerged out of a water-like chaos.

Matter and antimatter are mirror images of one another, the only difference being that they have opposite electric charges. This means that they can cancel each other out and return to pure energy (whatever that is!), and this can then create more matter and antimatter. It has been noted by scientists that if the Big Bang had resulted in a universe that was perfectly symmetrical, stable matter would not have emerged. And that means we would not be here. But more by accident than by design, as the temperature lowered and the speed of the jostling particles decreased, a small surplus of matter arose, and in virtually no time

at all what we now know as subatomic particles began to appear
and to form a kind of cosmic soup.

Subatomic particles are notoriously unstable. In 1964 two
scientists proposed the *quark* to explain what happens at the
subatomic level. They regard the quark as a fundamental con-
stituent of matter that cannot exist on its own but combines to
form composite particles called *hadrons*, in which the quarks are
held together by the strong nuclear force. Of the many kinds of
hadrons, the two most common types are those that form the
nuclei of atoms.

Also in 1964 Peter Higgs of Edinburgh postulated the existence of a carrier particle, or *boson*, which has the ability to endow
subatomic particles with mass. Since such a particle, if it exists,
would help to explain how mass came to exist in the universe, it
has come to be known popularly as 'the god particle'. To test this
theory, a very high-energy particle collider began to be built in
Switzerland in 2010, and in July 2012 an experiment confirmed
the existence of a previously unknown particle, though it is yet
to be shown whether or not it is the boson predicted by Higgs.
This is the way in which our knowledge is edging its way further
back towards the beginning.

We now know that no less than minutes after the Big Bang
there began what is called *nucleosynthesis*, a process in which
protons and neutrons united to form the nuclei of hydrogen
and helium. Yet incredible as it may seem, some four hundred
thousand years would pass before electrons combined with these
nuclei to form the atoms of hydrogen and helium. These are the
two simplest and most widespread stable elements of matter now
found throughout the known universe. After that, and over a
very much longer period, the gravitational force began to make
itself felt. The slightly denser regions of hydrogen began to attract one another and grow denser still. Thus were formed the
gas clouds out of which the stars and galaxies that we observe
today slowly evolved.

But can we observe all that the universe consists of? Just as cosmologists have made us aware of the mega-world that far transcends the macro-world we are familiar with, so they now have reason to suspect that the matter that forms the stars and galaxies of the mega-world may be only 20% of the universe as a whole. They now speak of *dark energy* and *dark matter*. Dark energy is a hypothetical form of energy that permeates all of space and has the tendency to increase the rate of the expansion of the universe. It may account for 73% of the total mass-energy in the cosmos. Dark matter is matter that emits neither light nor any form of electromagnetic radiation, and so cannot be detected by optical or radio astronomy. Its existence was inferred by the gravitational effect it apparently has on visible matter. In fact, ordinary matter—our main subject up to the present—may be only 4.6% of the total mass-energy of the observable universe, the rest being attributable to dark energy and dark matter. Clearly, then, our growing understanding of the universe, though still open-ended, is mind-bending. No wonder it is all very bewildering to non-specialists.

In this long process of cosmogenesis we see many examples of the two opposing tendencies I have referred to already. While the ever-expanding universe shows the spreading tendency, we observe the uniting principle first in the formation of hadrons, then of nuclei, and finally of atoms. This uniting tendency reveals itself not only in the formation of stable matter such as atoms, but also in the gravitational force that gathers the atoms together in clouds of gas. (We shall later see that biological evolution clearly exemplifies this very tendency.)

Over many eons gas clouds composed primarily of hydrogen along with some helium and trace amounts of more complex elements become denser and denser until eventually evolving into a star. Once the core of the star is sufficiently dense and hot, it becomes a nuclear furnace in which hydrogen is steadily converted into helium through the process of nuclear fusion.

The star's internal pressure (the spreading tendency) prevents it from collapsing further. This is the way that stars like our sun were born, and even now stars continue to be formed from gas clouds of interstellar particles.

Our sun is an average-size star. Three-quarters of the sun still consists of hydrogen but, because of its size, the hydrogen it contains is 150 times denser than water on Earth. The rest of the sun consists of the helium that this massive nuclear furnace is creating out of hydrogen by nuclear fusion. This process releases the enormous amounts of energy that the sun radiates in all directions, producing what we experience as sunlight and solar heat, yet what we see and feel is only a tiny portion of the energy that the sun is continuously generating.

Stars go through life cycles that depend on their size. Some, when they have exhausted all of their fuel, become white dwarfs; these are extremely dense, having a mass like that of the sun but a volume near that of Earth. They are very hot when they are formed but, having no further source of energy, gradually cool down. Stars the size of our sun expand into red giants as less and less hydrogen remains to be turned into helium. This expansion eventually results in degeneration and the recycling of a portion of its matter into the interstellar environment, where it will form new stars with a higher proportion of heavy elements.

Some massive stars become *supernovas*, a term coined in 1926. The process of becoming a supernova is so rapid that it can be observed as an 'event', and several such events have been observed during human history. A supernova is really a stellar explosion that may radiate as much energy over several weeks or months as the sun is expected to emit over its entire lifespan. It expels much of the star's material into space at great velocity. Complex processes that occur during the course of the explosion create heavy elements such as carbon, iron, magnesium and the rest of the 92 stable elements, comprising the stellar dust from which planets are eventually formed.

Just as gravitation (the uniting tendency) causes the gas clouds to become denser to form stars, it also causes the stars to gather together into galaxies. The word *galaxy* is of Greek origin and refers to the Milky Way, the name given by the ancients to the vast collection of stars to which our sun belongs. Some galaxies are elliptical in shape, some are spirals with long, curving arms, and some are irregular in shape. Modern telescopes even allow us to watch galaxies that are in the process of colliding with one another. Some dwarf galaxies contain as few as ten million (10,000,000) stars, while the giant galaxies may have as many as a hundred trillion (100,000,000,000,000). Since numbers of that magnitude are difficult for our minds to absorb, how much more staggering it is to learn that there are about 170 billion (170,000,000,000) galaxies in the observable universe!

At the centre of many galaxies scientists have detected what are called black holes. These are so called because the mass of each is compacted so densely that it absorbs everything in the vicinity, thus allowing nothing—not even light—to escape. A black hole is another example of a singularity—an object or place where the laws of physics no longer operate and both time and space lose their meaning. Einstein's theory of general relativity predicted that these phenomena would occur in the case of extremely massive stars, and cosmologists have now confirmed their presence. At the centre of many galaxies there may be black holes that account for up to 90% of their mass; indeed, strong evidence suggests the presence of a massive black hole at the centre of our galaxy, one with a mass four million times that of our solar system.

Late in the twentieth century some far-distant and very bright sources of light were discovered; they are now called *quasars*. The energy they emit in the form of light can be up to a thousand times that of the whole Milky Way, or that of a trillion suns. Cosmologists believe that they are powered by the accretion of material being attracted by a black hole, and that their apparent brightness comes from a corona surrounding the invisible black hole at the core. This corona may be regarded as the last despairing call for help by stars about to meet their demise. In even more colloquial language we could speak of a quasar as a bonfire of worn-out stars. It has been conjectured that in less than five billion years a quasar could form when the Andromeda Galaxy collides with our own Milky Way.

This is a brief description of cosmogenesis—the process by which the universe has been coming into being for more than thirteen billion years. This story is what we humans have so far been able to put together about the cosmos in which we find ourselves. It is a quick sketch of the truly awe-inspiring universe we can glimpse through our most powerful telescopes in the cloudless night sky—a universe whose magnitude is more than our minds can cope with and whose complexity takes our breath away as we learn about it. Its potential for expansion and ability to manifest itself in ever-more complex forms can only leave us speechless as we ponder it. Still more amazing is the mysterious fact that it has produced conscious beings like us who are able to study and contemplate it. The cosmos outstrips in magnitude and creative potential our traditional mental images of a creative God.

And all this has come about in the absence of any initial plan or initiating planner. Such subsequent designs as our human minds fasten upon with delight and awe have evolved more by chance than by any intent or plan. It's as if the universe blindly stumbles onward and creates designs by accident as it tries out all the possibilities that are contained in the basic energy of which it is composed. Of course, the universe is not itself a conscious entity, as God was generally imagined to be; it has no central self, and is quite indifferent to all of its own cosmic creations and events. As the astronomer James Jeans said in his book *The Mysterious Universe* (1930), "The universe is mostly empty space, so cold that life is impossible out there, while most of the objects in space are so hot that they make life equally impossible."

After more than thirteen billion years the universe still continues on its mind-boggling way. And even our attempts to penetrate its past can yield only very general answers, for we are encountering the universe at the mega-level, a level far beyond the macro-level at which we live. To learn more of what the self-creating universe has brought forth at the macro- and the micro-levels, we are confined to our cosmic home, planet Earth. Whether there are other planets with stories like ours, we do not know. Indeed, such are the distances that separate us from all other places where life might exist, it is unlikely that we shall ever learn the story of any one of them in anything like the degree to which we are coming to know the story of Earth. And yet the universe contains so many stars and galaxies that it is reasonable to assume that there must be other planets with stories like ours. It is within the ever-unfolding story of the immense mega-level space-time continuum that we now turn to the story of the planet on which we live. I shall refer to this story as *geogenesis*—the coming into being of Earth.

# Geogenesis

## The Emergence of the Earth

### From 4,500,000,000 Years Ago

For a very long time no suitable material existed to form a planet like Earth. First, the subatomic primal energy had to be transformed into stable matter, and the first such element, as we have seen, was hydrogen. Then the hydrogen had to gather into clouds and become dense enough to create the nuclear furnaces necessary to transmute hydrogen into such elements as helium. Only when some of these stars became supernovas (super-nuclear furnaces) could they produce the weightier stable elements—the kind of matter from which planets could be formed.

That is why it took the universe so long to bring forth planets, for indeed they are made of stardust—the stuff created and left behind by the explosion of worn-out stars. Thus a great deal of cosmic time necessarily preceded our planet's coming into being. Just as Rome wasn't built in a day, Earth wasn't built in a billion years. In fact, it took nine billion years for the universe to produce the material from which Earth came into being. Whether in some other part of the universe there are any planets exactly matching Earth we do not know, but it has recently been discovered that there are many other planets in our galaxy.

Various theories have been offered to explain how the stardust from a supernova was slowly gathered together by gravity to form our planet, but the most likely one proposes that, about five billion years ago, a vast cloud of gas stretching across many light years gradually formed a hot spinning disc called a solar nebula. As gravity caused some of this gas and other matter to grow denser, our sun was formed at its centre about 4.6 billion years ago. The heat and density of the sun was sufficient to ignite spontaneous fusion, and thus the sun was turned into the nuclear furnace that it remains to this day. The resulting solar wind then blew away from the sun all the remaining material in the gas cloud, including the heavier elements produced by the supernovas.

However, gravitation prevented this surplus material from wholly escaping into space and pulled it into orbit around the sun, where over many eons it coalesced into the millions of asteroids and eight planets that, together with the sun, comprise our current solar system. Meteorites that continue to hit Earth's surface show by their radioactive dating that they were formed about this same time, namely 4.57 billion years ago, and we know that early in Earth's existence such extraterrestrial objects inflicted a heavy bombardment on its surface.

The eight planets are usually divided into two types. The large, outer, low-density giants—Jupiter, Saturn, Uranus and Neptune—consist of gas, while the inner four—Mercury, Venus, Earth and Mars—are smaller and solid. Six of the planets are orbited by one or more moons (or *satellites*). In addition, there are five dwarf planets and possibly as many as a trillion comets. A comet is a conglomerate of ice, dust and small rocks that may vary in size from a few hundred metres to tens of kilometres across. It differs from an asteroid in that it has a coma or tail that is caused by the solar wind as it rounds the sun in its orbit. Thus, our planet has plenty of company in the solar system that slowly emerged out of the original solar nebula. To a dispassionate observer, Earth is simply one significant example of the universe's trillions of solar satellites.

Further, a theory known as the giant impact hypothesis proposes that about 120 million years after the formation of the planets (some 4.45 billion years ago) another proto-planet somewhat smaller than Mars collided with the primordial Earth and deposited its own heavier metallic core onto its surface (from where it sank into the Earth's core), while the rest of it glanced off to orbit around Earth as its moon. This would explain the moon's abnormal composition, for it is both less dense than Earth and more spherical. If this scenario is true, the impact had further important consequences for Earth: it released such an enormous amount of energy that it not only caused both Earth and moon to be completely molten for a time, but it also created the axial tilt that is responsible for Earth's seasons. (Just imagine how monotonous our lives would be without the changing seasons! In any case, without them life would not have evolved in the way it did.) This hypothetical impact, if indeed it led to the creation of Earth's satellite, may also have sped up Earth's rotation. By contrast, the moon rotates on its axis only once a month, exactly the same time as it completes its orbit round Earth, and that is why we never see the other side of the moon.

Thus it was that for a very long time the surface of Earth was so hot as to be wholly molten; only as it began to cool over its first billion years did it release massive amounts of carbon dioxide, steam, ammonia and methane. The heavier elements such as iron were drawn towards the centre of the planet, where they later became the source of Earth's magnetic field. A crust, largely of basalt, began to form on Earth's surface about four billion years ago, and the resulting slow cooling process allowed the steam to condense into the water that eventually produced our seas. The original solar nebula may have left an early atmosphere of such light elements as hydrogen and helium, but if so, the solar wind and Earth's heat would have driven it off. It was to be a long time before Earth would possess its current atmosphere, consisting of 78% nitrogen and 21% oxygen. The primordial Earth, however, was completely lacking in oxygen, so necessary to our kind of life.

At this point it may be useful to describe the basic layers of which Earth is composed. Most of it is the *barysphere* ('sphere of heavy material'), consisting of the dense central core and the surrounding ductile mantle—the former being solid, the latter liquid. Iron and nickel are dominant at the centre of the barysphere, but the mantle contains all the other elements in varying quantities. Surrounding the barysphere is a crust or shell now called the *lithosphere* ('sphere of stone'), which includes the outer brittle mantle and outermost crust and is only about one hundred kilometres thick. Stretching over the majority of the lithosphere is the *hydrosphere* (the oceans) while the whole globe is surrounded by the *atmosphere*. The hydrosphere and atmosphere constitute a relatively thin envelope surrounding Earth. Yet it was within this envelope that there gradually evolved what has been called the *biosphere*, or sphere of life.

Unlike the other planets in our solar system, only Earth can be called a watery planet: indeed, the oceans cover 71% of its surface and contain 97% of Earth's water. This hydrosphere began to form about four billion years ago, and while much of it resulted from the condensation of water vapour produced by the early volcanic activity, the total volume is such that scientists now believe a considerable portion must have come from the impact of comets and proto-planets that have continually bombarded the planet.

Water, since it is essential for the evolution of life, is Earth's most valuable natural asset. It exists widely in three forms—solid, liquid and gas—but also less obviously as molecular water in many compounds, both organic and crystalline. Water is also the medium that drives our weather patterns with its continuous cycle of evaporation and precipitation, a process that has the added effect of continually purifying it. So vital to life is the presence of water that it is not surprising that the biblical story of origins points back to a watery chaos out of which an ordered universe of sky and Earth was created. While it has become absurd even to suggest that the space-time cosmos was created out of a "watery" chaos, it is nevertheless true that our human world was made possible by the presence of the hydrosphere.

And did the hydrosphere once cover the whole Earth? Possibly it did! That renders even more apt and serendipitous the biblical account's assertion that God commanded the dry land to appear. From the evidence still extant in its rocks, geologists and geophysicists are attempting to piece together the early story of the lithosphere. It seems that the first crust that formed in the cooling process has now largely disappeared, either buried by subsequent layers or subducted and assimilated back into Earth's mantle. The oldest known rocks—found in small areas of Greenland, western Australia, southern India, South Africa and Antarctica—come from the very first continent, named Ur, which formed some three to four billion years ago.

Beneath the cool, brittle lithosphere lies a mobile, ductile, hotter and thicker layer of Earth's mantle called the *asthenosphere* (weak sphere). The lithosphere, being much more rigid, has long been broken up into tectonic plates that have both a higher strength and a lower density than the underlying asthenosphere. Fifteen major tectonic plates, along with many minor ones, appear to float or ride over the asthenosphere. Over long periods of time they collide with one another, exerting enormous surface pressure and causing earthquakes and volcanic eruptions. They have pushed up mountain ranges such as the New Zealand Alps and the Himalayas that now divide India from the Tibetan plateau. Our present knowledge of tectonic plates (phenomena first postulated as recently as the mid-1960s) helps explain, among other things, why the coastline of West Africa seems to match that of the east coast of South America, while a similar fit frames the vast expanse of the Pacific. The latter congruence becomes even more apparent if, rather than present coastlines, the submerged continental shelves are compared.

During the last three billion years, the continents formed and have been drifting about, changing shape and size. For instance, about 750 million years ago a land mass named Rodinia broke up into various continental fragments, which then coalesced about 550 million years ago into a single landmass known as Pannotia. This then split up into two large continents known as Laurasia and Gondwana. Subsequently, these two continents regrouped to form Pangaea about 275 million years ago, only to break up again! In time, Gondwana separated into the current southern continental landmasses. The parting of Africa and South America about 180 million years ago led to the formation of the Atlantic Ocean. Africa, India and South America were the first to break away from Gondwana, about 125 million years ago, and India collided with Asia about 45 million years ago. Zealandia drifted away from eastern Gondwana, causing the formation of the Tasman Sea floor (between 83 and 53 million years ago). Finally, when Antarctica moved over the South Pole Australia broke away from it, and South America became attached to North America. And the wrenching and colliding goes on; Earth's crust is relentlessly changing as sea floors spread and continents drift.

The atmosphere surrounding the hydrosphere and continents is an envelope of gases that has no boundary but just fades into space. However, 97% of it lies within thirty kilometres of the surface of Earth, and that makes it about the same thickness as the continental crust. How did it come into being? During the first five hundred million years there emerged an early atmosphere consisting of water vapour and gases—primarily hydrogen, nitrogen, methane and carbon dioxide. What it lacked, compared with that of today, was oxygen, the arrival of which came much later and largely as the result of biological activity.

To complete the story of the coming into being of our Earth, we must now include a preliminary description of the story of planetary life, the topic of the next chapter. The two stories overlap because what makes our Earth such an interesting and delightful habitat for humankind—a veritable Garden of Eden, in fact—is what took place on its surface during the second half of Earth's story. If not for these events, our planet would have differed little from Mars or Mercury; Earth would have been no more than an enlarged lifeless moon endlessly circling the sun, heating up and cooling down every twenty-four hours.

Before the planet could provide the conditions necessary for the survival of today's common forms of life, many changes still had to occur. Since the Earth used to rotate faster on its axis than it does today, days were shorter; and because the moon was closer to Earth, the tides were formidably stronger. Still other changes resulted from subtle but important interactions among the surface layers of Earth—the lithosphere, the hydrosphere and the atmosphere. And when the earliest forms of life appeared, simple though they were, energy coming from the sun in the form of heat and light provided essential advantages.

Scientists now conclude from the geological evidence that some two billion years after the planet formed, aquatic organisms called blue-green algae (not the first forms of life) used a solar energy conversion process called *photosynthesis* to split molecules of water and carbon dioxide in such a way as to release oxygen and allow them to recombine to form other organic compounds. This is how oxygen first came to enter the atmosphere and gradually accumulate, until some 2.5 billion years ago when the atmosphere contained enough free oxygen to support most of the later forms of life. It is fascinating that these relatively primitive forms of life were instrumental in providing the necessary conditions for the emergence of the myriad later and more complex organisms we now know.

Oxygen is a very powerful gas, as is shown by how quickly it acts on iron to form rust. Indeed, its arrival in the atmosphere caused an ecological disaster by destroying a number of primitive organisms. On the other hand, the increased level of oxygen in the atmosphere led to a decrease in the amount of carbon dioxide and set the stage for the development of increasingly higher forms of life.

A further result of the arrival of oxygen, with yet more fortunate results, is that high up in the atmosphere the $O_2$ molecules of oxygen split into $O_1$ atoms of oxygen, which then combined with $O_2$ molecules to form the $O_3$ oxygen we call ozone. The thin layer of ozone that now exists in the upper atmosphere was to prove vital for the evolution of life, since it is very effective in absorbing forms of solar radiation that are dangerous to life. Thus, while one sort of oxygen has become essential for most forms of earthly life, another has provided for its preservation by acting as a shield that protects the planet from the sun's nuclear radiation.

Indeed, not until a sufficient layer of ozone formed in the upper atmosphere was it possible for the evolution of terrestrial animals to take place. The necessary amount of ozone was reached about six hundred million years ago, prior to which life was largely restricted to the ocean. It was the presence of ozone that enabled some organisms to move onto the land and evolve there.

A further essential development, known as *pedogenesis*, was the formation of soil through complex bacterial and chemical processes. Soil has been as essential for the evolution of plant life, as plant life in turn has been essential for the evolution of most animals. The study of soil and its adequate care has recently become a very important branch of science, but its original formation was unplanned and exceedingly slow. Pedogenesis is simply another example of the many chance events and processes on which the evolution of life depends.

And now we are ready to turn to the account of how life evolved within the thin envelope of the hydrosphere and atmosphere, the next chapter of the story of our planet. As frequent earthquakes and volcanic eruptions remind us, the process of geogenesis goes on. And geogenesis remains but a minor episode in the life of the cosmos. Cosmologists predict that in some four billion years Earth will be swallowed up by the sun, which by that time will be entering its dying phase by expanding into a red giant.

While we may not lose any sleep over that far-off cosmic event, we do need to remember that Earth remains vulnerable to whatever may come wandering into our solar system from outer space; in fact, we are bombarded daily with tiny meteors that we see as shooting stars. Most of them do no harm since they are burnt up in the atmosphere, but very occasionally we encounter larger ones, like the one that created the moon and the one that brought an end to the dinosaurs only 65.4 million years ago. Earth was formed more by accident than by any design, and it remains subject to such chance cosmic events. It does not exist for any particular reason, even though from our subjective viewpoint it may seem that Earth exists for the purpose of bringing forth life in general and of creating us in particular. It is to that coming-into-being of planetary life that we now turn.

# CHAPTER 3

# Biogenesis

## The Emergence of
## Life on Earth

*From 3,000,000,000 Years Ago*

For nearly a billion years the volcanic surface of Earth was so hot that it could not sustain any life at all. This earliest geological eon has been appropriately called the Hadean, from the Greek *Hades*. For a very long time Earth was so hellish as to be lifeless, yet today the oceans swarm with aquatic creatures, and myriad birds, animals and insects inhabit the land surface and the air above—despite the geological evidence that the majority of species that have ever lived on Earth or in the sea are now extinct! Where did they all come from? It was during the second billion years of Earth's history, called the Archean eon, that very simple forms of life first made an appearance.

Some have supposed that these life forms originated elsewhere and arrived here on a comet. This would mean that all planetary life evolved out of such a visitation from outer space. But such a theory does nothing to indicate how life originated in the cosmos; it simply moves that genesis one step further into space. Wherever the transition occurred, life must have evolved out of non-living physical substance; accordingly, it is preferable to look for the transition here on Earth where we have some evidence to work with.

The question of how even the simplest forms of life could possibly evolve out of non-living material seems to some people to require a miracle, a supernatural intervention by an external force. That is why creationists feel their argument for the existence of a creative God is so compelling. But to state that all living creatures were created by a creator deity is *no explanation at all* for the origin of life; it is simply a *declaration* that shifts the seat of the mystery to a point hidden behind a divine screen that we are forbidden to penetrate. This traditional explanation of the origin of life made some sense in the ancient cultural world of Babylonia in which it emerged, and in a later chapter we shall discuss how the concept of God was born in the collective mind of the ancients and how until quite recently it played an important role as a powerful unifying symbol. To perpetuate such an interpretation today is to resort to a sleight-of-hand argument that should be recognised for what it is.

Of course, if *miracle* is assigned its original meaning of 'something to be wondered at' rather than 'a divine suspension of natural law', then the emergence of life was indeed a miracle. The story of the evolution of life from its simplest beginnings to the emergence of the human species is truly awe-inspiring.

Our problem is that, having grown up surrounded by living creatures, we take life in general so much for granted that we no longer appreciate how truly miraculous life is. Yet have we not already found from our discussion of cosmogenesis that the expanding universe itself is enough to fill our minds with wonder? When we begin to study the origin of life within the context of this mind-boggling universe, we can see that biological emergence is fully consistent with the very nature of this continually developing universe.

However impossible the leap from non-life to life may seem, it should be recognised as but a further example of what has been a succession of seemingly miraculous leaps within the evolving cosmos. We have already observed that early on in cosmogenesis, hydrogen began to be turned into helium. We speak of that as a nuclear transformation, for it required a change in the nucleus of the atom, yet that constituted a very significant leap. The reason why the mediaeval alchemists were never able to turn base metals into gold was that they did not understand that it requires a leap of nuclear proportions at the subatomic level to turn any one of the ninety-two elements into another.

As we now know, the higher a particular element stands in the atomic scale, the more complex its unique collection of sub-atomic particles. Further, since molecules contain two or more atoms, they are inherently more complex than atoms, and *mega-molecules*—so-called because they contain thousands of atoms—are incredibly more complex. When we reach the simplest unit of life, the cell, we find a further great leap in complexity; not only does it consist of many mega-molecules, but they engage in chemical activity with one another in such a way as to form a living, complex whole.

Teilhard de Chardin usefully coined the term *complexifica-tion* to describe this inherent tendency of the universe to join simple units together into increasingly complex ones. Our study of physical matter shows us that the universe is continu-ally organising itself into ever-more complex units, and it does so without any external aid or intervention! In short, the uni-verse is not a static reality but a creative process. Teilhard called this extraordinary organisational principle of nature the *law of complexification*.

In 1926 Jan Smuts of South Africa coined the term *holism* for this phenomenon in his book *Holism and Evolution*. He defined it as "the proneness of the universe to form wholes". His term means much the same as Teilhard's, and similarly represents the key to the evolving universe in general and to the evolution of planetary life in particular. In other words, it is just as *natural* for the universe to bring life out of non-life as it is for hydrogen to be transformed into helium. The universe brings forth life through its natural proneness to form wholes.

The reason it at first seems incredible that life could be cre-ated out of non-life is that we fail to appreciate how wide the spectrum of living forms is—from highly developed plants and animals to the other extreme of viruses, which, as we shall see, can hardly be said to be alive. Our familiar plants and animals were far from being the first forms of life to appear on the earthly scene; indeed, they are very late and complex forms of life whose emergence was preceded by an extremely long period in which planetary life existed only at the microscopic level.

The scientific study of how biological life emerged out of inorganic matter through natural processes is known as *abiogenesis*. For an entity to be classified as 'alive' it must be endowed with two essential capabilities: First, it must possess a set of chemical actions and reactions adequate to produce growth. These processes, collectively called *metabolism* and carried out by catalysts known as enzymes, perform two functions that are called *catabolism* and *anabolism*; the former is the breaking down of the organic matter on which the organism feeds, and the latter is the construction of such necessary components as proteins and nucleic acids. Second, it must possess the power to replicate itself and thus perpetuate the species. The simplest forms of life do this by dividing into two. The more complex forms do it by the coupling of male and female genders.

The smallest organic unit that possesses these two capabilities is the cell, which is therefore often called the building block of life and is the basis of all life, for every living creature is a community of cells that cooperate in amazing ways. We do not yet know how the first cell came to be formed, nor has anyone been able to synthesise the basic chemical components into a living cell. The twentieth-century discovery of the virus seemed at first to have implications for understanding the origin of life, for although a hundred times smaller than the smallest cell it does have the power to replicate itself. But opinions now differ as to whether the virus is a form of life at all, for it does not have its own metabolism and can only replicate itself inside a cell. Thus the virus is restricted to a parasitical existence, dependent on other forms of life. Although it does not explain how life originated and may be described as an organism at the edge of life, it nevertheless helps us to appreciate the microscopic character of the earliest forms of life.

That the cell is an extremely minute entity can be seen by the fact that the human body contains some hundred trillion of them. Yet in spite of its microscopic size, the cell is also extremely complex. In fact it is a veritable chemical factory in which many chemical reactions are continuously taking place. Within this factory are simple chemical molecules like sugars, complex molecules like proteins and nucleic acids, and the enzymes that aid the various chemical processes. Just as our bodies are protected by a skin, so the cell is covered with a membrane that provides protection from the external world by guarding the cell's nucleus and the various compartments where the metabolic activities take place. The membrane carries out the very important function of determining what should be allowed to enter the cell and what should not.

Inside the nucleus of the cell is the now well-known organic unit known as DNA. This is a veritable library of information, containing the instruction manual for the development and function of each organism. This genetic code, as it is also called, takes the form of a double helix of extraordinary length in which a few basic elements of matter are ordered in a particular way. The combinations of these elements are known as genes, and these determine the perpetuation of specific features when the cell propagates itself by dividing into two new cells.

Before the division of a cell takes place, the DNA is copied so that each of the resulting two cells will inherit the DNA sequence. It is hardly surprising that in replicating a sequence as complex as the genetic code mistakes occasionally take place, either during the life of the cell or in the copying process. These mutations, as they are called, can lead to subtle differences in the nature of the resulting cell and the way it operates. The changes may be beneficial or harmful, and while most fall into the latter class and lead to the death of the individual, the few beneficial ones enable the cell to interact more effectively with its environment and thus have great survival value. Today, when humanly designed genetic modification has become a hotly debated issue, this natural process of mutation may be regarded as genetic modification by chance.

Thus, as DNA is endlessly copied and passed on, it very slowly diversifies. And while neither Charles Darwin nor his contemporaries had the slightest knowledge of it, DNA is the key to the theory of evolution by natural selection that Darwin first enunciated. The mutations that open the door to natural selection clearly demonstrate that it is not the operation of some purposeful force but rather chance events that have led to the amazing diversification of life into its innumerable species. Further, the study of DNA has shown that all living organisms on the planet have descended from a common origin.

But how did such a complex entity as the first living cell arrive on the scene? In the nineteenth century Charles Darwin suggested that life may have begun in a "warm little pond, with all sorts of ammonia and phosphoric salts, lights, heat, electricity, etc, present, so that a protein compound was chemically formed ready to undergo still more complex changes". This rather vague guess may yet turn out to be not far away from the mark, for at White Island, an active volcano off the coast of New Zealand, scientists have discovered primordial forms of life in a very hot lake.

One thing we can say with certainty is that such a complex biological building block did not suddenly come out of the blue. The first cell took a very long time to evolve. We know from geological evidence in fossils that the earliest signs of life are to be found in the Archean eon, which stretched from 3.8 to 2.5 billion years ago. These primitive forms of life may have emerged in the ocean near vents in the Earth's crust, where greater heat and the presence of unusual mega-molecules triggered a planet-altering singularity.

But what sort of microorganisms were they that left deposits in the fossils? Here we need to distinguish between different types of cells. The one briefly described above, which is found in all plant and animal life, is called the *eukaryote*. But there are also the more heat-loving ones known as the *prokaryotes*, these were probably the very first cells to be formed.

We have known of the prokaryotes for a century or two, and we call them bacteria. They are still with us everywhere and are extremely numerous; a millilitre of fresh water contains a million of them, and forty million can be found in a gram of soil. Large numbers of them live on the human skin and in the human gut. They have always played a crucial role in life processes, for they recycle nutrients. And while some are beneficial and most are rendered harmless by the body's immune system, a few of them can be fatal by causing such infectious diseases as cholera, leprosy, tuberculosis and bubonic plague.

These unicellular organisms seem to have been a necessary precursor to the more complex eukaryotes, for they have no nucleus and are a thousand times smaller. It is thought that the eukaryotes may have evolved from a symbiotic community of prokaryotes, though it is still uncertain when and how such a development took place. Assuming that it did, this would be another example of the principle of complexification.

It has been the universe's continual tendency towards complexification, associated with the mechanisms of mutation and natural selection, that has resulted in the great diversification of life into countless species. Just as Archean mega-molecules became organically united to form living cells—first prokaryotes and then eukaryotes—so in the Proterozoic eon did cells unite to form living multicellular organisms. Stretching from 2.5 billion to about 542 million years ago, this eon took its name from the Greek for 'earlier life', for it was once thought to be the eon during which life began.

Emerging by virtue of their complexity, the multicellular organisms understandably increased the opportunity for chance mutations to occur. This meant that the diversification of life forms began to accelerate and the rate at which life evolved began to increase. We may therefore extrapolate from Teilhard's law of complexification what might be called the *law of complexity-diversity-acceleration*, for the more complex organisms become, the faster they diversify into new species.

In the Proterozoic eon all life was still confined to the ocean. The multicellular organisms that began to emerge during this period of diversification were the forerunners of what became the five or six so-called kingdoms into which life has been catalogued. It was Aristotle (384–322 BCE) who first conceived the idea of dividing life into kingdoms, and he discerned two: plants and animals. Modern scientists have widened the number of kingdoms to six: bacteria, protozoa (e.g. Amoeba), chromista (e.g. diatoms, mildew, kelp), plants, fungi and animals. To be sure, plants diverged from animals at a very early stage, but we must also remember that many early life forms were very different from their modern descendants. Today we find it more convenient to think in terms of such categories as algae, kelp, jellyfish and flat sea worms.

The Proterozoic eon came to an end about 542 million years ago. At this point in the story of how life came into being we may find it a suitable time to pause and reflect on how far we have come in the story of cosmogenesis. First, recall that thirteen billion years have passed since the Big Bang. Then think of the mind-boggling contrast between the unbelievable speed with which the cosmos burst forth in the first few seconds and the slower-than-glacial pace with which the stars, galaxies and planets came into being. Only after all of that did life finally appear on one of them. Even then the evolution and diversification of earthly life has stretched out over three billion years. Evolution takes time—lots of it.

The vastness of time and space is the essential context for understanding both the origin of life and its subsequent gradual diversification into kingdoms, families, genera and species. Since a radical threshold of change has to be crossed for organic life to emerge from inorganic matter, it is not wholly surprising that the coming into being of life was a long and drawn-out process. It all started with the prokaryotes. Just stop and think—for more than one billion years the only life on Earth consisted of microscopic bacteria. This leads to a surprising and perhaps unsettling conclusion: far from being descendants of the mythical Adam of biblical fame, we must look to the oft-despised bacterium as our ultimate ancestor.

After the prokaryotes were joined by the more complex eukaryotes, it required the further two billion years of the Proterozoic eon for the holistic tendency of the universe to bring the eukaryotes together and form multicellular organisms. And after this nearly unimaginable expanse of time, life was still confined to the oceans, in forms quite unfamiliar to us. By the end of the Proterozoic eon, Earth had existed for nearly four billion years, but still no plants or animals of the kinds we are familiar with had yet appeared. Indeed, there was no life at all on the land. How life in the oceans evolved further and at last not only invaded the land but eventually produced us is the story to which we now turn.

# CHAPTER 4

# Anthropogenesis

## The Coming into Being
## of Humankind

### *From 570,000,000 Years Ago*

The story of Earth (geogenesis) is closely entwined with the story of life (biogenesis). This is so because all known life came from the Earth (all living creatures are 'dust from dust') and because the survival and continuing evolution of all life forms is shaped by the conditions of Earth. But it is equally true that conditions on the planet have been and continue to be changed by the life it has produced. That has become clear quite recently with the advent of global warming. In short, a symbiotic relationship exists between the ever-changing Earth and the evolution of life to which it has given rise. This relationship has become the subject matter for the new science of ecology.

With regard to this relationship, there are two aspects of earthly conditions that need to be noted here as we proceed to examine the period of biogenesis that eventually produced the human species. First, it must be recalled that six hundred million years ago most of Earth's surface was covered with a vast expanse of water broken by a single land mass, now known as Pannotia.

Second, we now know that Earth has experienced changes in global temperatures so severe that they have affected and even halted the evolution of life. The planet has experienced five major ice ages, the first of which occurred more than two billion years ago during the early Proterozoic eon. The second, the largest in the last billion years, came at the end of the Proterozoic, at which time glacial ice sheets reached from the poles to the equator. The third and fourth occurred from 460 to 420 million years ago and 350 to 260 million years ago respectively. A fifth began more than 2 million years ago, since which time ice sheets have advanced and retreated over roughly 50,000-year periods.

Between the ice ages there have been relatively cold eras known as *glacial periods* and warmer ones termed *interglacials*. During the glacials, masses of sea ice extended far outward from the poles, removing such large amounts of water from the ocean that sea levels were significantly lowered. Perhaps this was a factor in the eventual invasion of land by aquatic life forms. The last glacial period ended about ten thousand years ago, at which time we entered the current interglacial epoch known as the Holocene. Whatever the status of the planet, all forms of life have been obliged either to adapt to these changing climatic conditions or else become extinct.

In the late Proterozoic eon, around six hundred million years ago, our planet shivered in the grip of a major ice age for as long as thirty million years, but as it slowly warmed up Earth experienced a great surge in the creative process. Just as each annual spring ushers in a period of rapid growth with resulting flowering and fruiting, so this leap forward was a geological spring on the grand scale, and it burst forth with an astronomic profusion of new varieties and species of life forms. This flowering, called the Phanerozoic eon, extends up to the present day. The name *Phanerozoic*, from the Greek for 'the appearing of life', was appropriately chosen, for only in this last half billion years of Earth's existence has life on the planet became clearly visible on the macro level. Thus, after a very slow beginning that lasted some three billion years, the creative process that had brought forth the first signs of life began to speed up. So comparatively rapid was this acceleration that from now on in our story we must deal in terms of millions instead of billions of years—while remembering that a million years is still a very long time!

Between about 540 and 490 million years ago, the diversification taking place among the multicellular organisms in the sea began to accelerate so fast that it has been called the Cambrian Explosion (named after the geological era during which it occurred). Despite the possible implication of the term *explosion*, it was far from being a destructive event. Rather, the word reflects the dramatic acceleration of biodiversity in a hyperbolic curve from a few to several thousands of genera. This was an extreme example of what we have called the law of complexity-diversity-acceleration. Indeed, from this time onward the story of the evolution of life becomes so complex that for our purposes it must suffice to sketch only the main outlines of the changes taking place in life forms.

We need also to remember that this most prolific outburst of life on Earth came about by the operation of natural forces and not because of the purposeful guidance of some unseen hand. The undeniable truth of this observation is attested by the fact that the headlong acceleration in the evolution and diversification of life was shockingly interrupted by a succession of mass extinctions. In fact, as a result of five such major interruptions to the evolution of life, scientists now estimate that 99% of all living species brought forth by the creative process are already extinct. Ironically, however, these extinctions sometimes had the effect of accelerating the diversification of species, since the extinction of a dominant group suddenly opened up new opportunities that allowed one or more of the surviving minor species to spread, flourish and diversify.

We may note in passing that these extinctions are strangely reminiscent of the biblical story of Noah and the Great Flood that destroyed all life except for those animals and one human family that were saved in the ark. That legend may well have sprung from memories of successions of disastrous floods that used to occur in the wide, flat valley of the Tigris and Euphrates rivers. Cultural traditions of those days (as we shall see in later chapters) always presupposed a reason for natural disasters, and typically interpreted them as acts of divine judgment. But today we are led by our accumulation of knowledge to conclude that earthly events are due to the operation of natural forces inherent in the physical world. Even though these forces have a natural tendency towards increasing complexity, they can hardly be said to have purposefully planned the life they brought forth—nor can we imagine they have any interest in preserving it.

The Phanerozoic eon, in which life as we know it really took off, has been divided into three sections: the Paleozoic era (541–252 million years ago), the Mesozoic era (252–66 million years ago) and the Cenozoic era (66 million years ago to the present). The sharp lines of division among the three eras demonstrate the relatively sudden ends of the first two as the result of two cataclysmic destructions of life noted above.

The first era, the Paleozoic, gave rise to some amazing new developments. As the ice-bound planet slowly warmed, it began to swarm with new life in amazingly diverse forms. It was as if nature was experimenting with all the possibilities at its disposal. The success of each new life form depended on how well it was suited to its environment and how effectively it coped with enemies and competitors.

We noted in the last chapter that Proterozoic life forms were already dividing into the plant and animal kingdoms. Other broad divisions now began to emerge in the animal kingdom, particularly that between vertebrates and invertebrates. The vertebrates owe their origin to the chordates, primitive sea worms with a dorsal nerve cord. They became fish, with structures marked by a spinal cord and internal skeleton of bone. By contrast, the invertebrates developed external skeletons in the form of a shell that provided durability and protection. Chief among the later invertebrates are the arthropods (the term means 'jointed feet'), a group comprised of all those creatures later classified as insects and crustaceans.

Life had been confined to the ocean for three billion years, but now organisms began to venture onto the lifeless land mass of Pannotia. First to arrive was plant life, which quickly adjusted to the conditions in which it found itself and rapidly began to diversify. By three hundred million years ago there were towering rain forests in tropical areas. By invading the previously barren land, plant life was unintentionally preparing the way for animal life to follow by providing the nutrients these highly mobile life forms would need.

Next to establish themselves on dry land were the arthropods (mainly the insects), and evidence of their presence has been dated to four hundred million years ago. Their structure made them well suited to adapt to land conditions, for while their external skeletons protected them from rapid drying and the relentless pull of gravity, their jointed appendages provided means of both locomotion and food-gathering.

The vertebrates, however, could not invade the land until they had undergone a number of necessary modifications—changes that took place in the Devonian period, from 419–359 million years ago. This has often been called the Age of the Fishes, because it marked a time of great diversification in these aquatic forms. Some began to develop lungs, a change that was to prove essential before animal life could exist on dry land. About 340 million years ago lobe-finned fish began to take on an amphibian existence like that of present-day frogs. Not only were they air-breathers but they began to use their fins for locomotion on land, and eventually they evolved into tetrapods, four-limbed animals with a spinal column. Their scales became the scales of reptiles, which much later turned into birds' feathers. Once these ancient amphibians had fully adapted to conditions on land, they became the ancestors of all subsequent animals—reptiles, birds and mammals. It should also be noted that this process of evolutionary adaptation from sea to land is similar to a later reverse 'migration' that occurred when some land animals found it advantageous to return to the ocean and thereby became the ancestors of whales, dolphins and seals, and when some birds took to the ocean to become penguins.

At this stage the tetrapods spreading over the land were small lizard-like creatures that eventually divided into two main groups: the sauropsids (from which came the reptiles and the birds) and the synapsids (from which came the mammals). About 300 million years ago a group of reptiles diverged from the rest; they are called the diapsids for their distinguishing mark of two holes on each side of the skull. It is from these that the dinosaurs arose, as well as the reptiles that have survived to this day—crocodiles, lizards, snakes and the New Zealand tuatara, which still looks like a miniature dinosaur and is the only living reptile with a diapsid skull. Thus, by the end of the Paleozoic era the land had become largely dominated by air-breathing, egg-laying reptiles. Most of them were cold-blooded, but a few, known as the therapsids, were warm-blooded and were destined to become the ancestors of the mammals.

About 250 million years ago the Paleozoic era was brought to a sudden end with a catastrophic global event that has been called the Great Dying or the 'mother of all extinctions'. Up to 96% of all marine species and 70% of vertebrate species on land became extinct, and this was accompanied by a mass extinction of insects. It is not known whether this catastrophe occurred as a swift succession of phases due to rapid environmental change or resulted from one sudden volcanic event, but—though the restoration of marine life may have been more rapid—it took land animals more than 30 million years to recover.

That great extinction ushered in the Mesozoic era, during which the surviving species exploited new opportunities to multiply and diversify. The first to make their presence felt were the dinosaurs, a group of tetrapods already referred to as the diapsids. Indeed, the Mesozoic era is often called the Age of the Reptiles, for they came to dominate Earth during that time. Though popular imagination has focused on the size of such giants as *Tyrannosaurus rex*, the average size of the more than one thousand species of dinosaurs was of the order of us humans or even smaller.

About 200 million years ago the most advanced synapsids, now called the therapsids, separated from the main body of reptiles to become the ancestors of mammals. Equally important for later times was the evolution of the theropod dinosaurs around 160 million years ago. Their scales became feathers as they took to the air and became birds. The new opportunities afforded by being relatively free of competitors enabled them to multiply and flourish, as they continue to do to this day.

Unlike the reptiles from which they evolved, mammals have a constant body temperature. They nurture their young first by a gestation period in the womb and then by suckling them from the female breasts after they are born. Early mammals were small and shrew-like, and fed on insects, but it was in these new creatures that the neocortical region of the brain began its unique and crucially important evolution. Then, about 120 million years ago—well within the Age of the Dinosaurs—the mammals divided into the marsupials and the eutherians (placental mammals). Marsupials, of which 70% of the extant population is now found in Australia, are unique in that females give birth to their offspring at an earlier stage of fetal development and then care for them in a pouch. The extant evidence indicates that the first marsupials appeared a few million years earlier than the first eutherians. Although our knowledge of this early stage in the evolution of the mammals is still very sketchy, these developments and others had taken place well before the end of the Mesozoic era. Further, as early as 85 million years ago, mammals may have separated into two basic groups: the rodents, rabbits and shrews on the one hand, and on the other the primates, from whom came monkeys, apes and humans. Designation as primates evidently resulted from their having been judged to be mammals of the first or highest order!

The Mesozoic era came to a sudden end 66 million years ago when a meteorite of considerable size collided with Earth and led to such a dramatic change in environmental conditions that the result was yet another widespread extinction of life. About 75% of all species became extinct, including the majority of the dinosaurs. The most important survivors were the insects, the mammals and the birds, while the ecological vacuum left by the departure of the dinosaurs provided a niche in which the mammals could flourish as never before.

The third and final era of the Phanerozoic eon is called the Cenozoic (from the Greek for 'new life'). It certainly witnessed a great deal of new life, chiefly among the order of mammals, and therefore has been called the Age of the Mammals, just as the Mesozoic was the Age of the Reptiles and the Paleozoic the Age of the Fishes. Yet despite the ascendancy of mammals, twice as many new species of birds appeared, and snakes also greatly multiplied.

The mammals that survived this second great extinction did not continue as the few basic forms found at the beginning of the Cenozoic era, but evolved into the great variety now seen all around us or in any good zoo—and that says nothing of the many that have now become extinct. Strangely new and different are such domestic mammals as cats, dogs, sheep, cattle and horses; the less familiar elephants, camels, hippos and monkeys; and we must not forget those that returned to the ocean and became whales, dolphins and seals. But being chiefly concerned with our human ancestry, we shall now focus on the order of mammals known as the primates.

The primates were already becoming a distinct order during the Mesozoic era; indeed, they are among the oldest of all the surviving placentals. At first they lived in the trees, as many of them still do. All primates possess the ability to climb trees, but some, such as the great apes and baboons, eventually became more terrestrial. The primates developed larger brains than other mammals and came to rely more on stereoscopic vision than their sense of smell. Most are prehensile—that is, they have hands capable of grasping. And although primates take longer to reach maturity than other mammals of the same size, they live longer lives. At some stage the order of primates divided into a number of subgroups, but opinion is still undecided on the best way of understanding how this occurred.

The lemurs separated from the other primates some 60 million years ago. What are called the New World Monkeys (including spider monkeys and marmosets) split off about 40 million years ago, while the Old World Monkeys (such as baboons) became separate about 25 million years ago. About 260 species of monkeys exist today and most of them live arboreal lives, though some, like the baboons, have taken to the ground. Monkeys are generally recognised as quite intelligent and, unlike the apes, usually have tails.

Of those species known as the Great Apes or hominids, the gibbons became distinct about 18 million years ago, the orangutans about 14 million years ago, and the gorillas some 7 million years ago. It is only about 6 million years ago that our human ancestors separated from the chimpanzees, and our DNA is about 98.4% identical to theirs. Yet we need only compare a human and a chimpanzee to see what a difference that 1.6% of DNA can make! It is impossible to be sure what our common ancestor looked like, but they probably appeared more like today's chimpanzee than today's human. It seems certain that we humans are the ones who have changed the most.

Today, then, the chimps remain our nearest living relatives among the family of Great Apes. Yet between ten and five million years ago as many as 20 different species of Great Apes existed, all of which were nearer to us than the chimp, and all became extinct. One of them, now called *Australopithecus* (southern ape), spread throughout Eastern Africa between four and two million years ago. The fossil remains of one such individual (known as Lucy) were recently found in Africa and dated to 3.2 million years ago. Although *Australopithecus* had a brain size only about 35% that of modern humans, they were certainly bipedal, for they left footprints to show it. Bipedalism gave the hominids a great advantage over other mammals, for the ability to stand and walk upright freed their hands to grasp objects and carry food and their young. Though they cannot be certain, anthropologists believe that the genus *Homo* evolved from this ancestor.

Scholars generally agree that our ancestry can be traced back to Africa. Like the family of Great Apes before it, the genus *Homo* diversified over the last two million years (dating from the onset of the latest ice age) into several species, the earliest arising in Africa and others later in Eurasia. These species have been identified and named as successive fossil finds in modern times revealed evidence of their existence. As yet we have no definitive picture as to how these species are related, either to one another or to modern humans, for even when genetic mutation gave rise to a species closer to ours, the resulting two species could have coexisted for some time before the older one became extinct. For this and other reasons, much of the evidence is open to a variety of conclusions, and consequently the debate continues. It is sufficient for our story to name the most important ones in their probable chronological order.

The oldest species of the genus *Homo*, the first to appear after *Australopithecus*, has been named *Homo habilis* (handy-man) because of his ability to create and use primitive stone tools. Of all the known species of the genus *Homo*, *Homo habilis* was the least similar to modern humans. He was short, with disproportionately long arms, and had a brain capacity about half that of ours. He may have been the ancestor of the more advanced *Homo erectus* (now equated with *Homo ergaster*), though some think the two species coexisted in Africa for as long as five hundred thousand years. The best current evidence suggests that *Homo habilis* lived from about 2.5 to 1.6 million years ago and *Homo erectus* from about 1.5 to 0.5 million years ago.

*Homo erectus* was so named because he was thought at the time to be the first human ancestor to walk fully upright. He developed a larger brain than his ancestors and made stone tools, and may have used fire to cook his meat. Some think he had a well-developed sense of self and some language. It is probable that *Homo erectus* is the common ancestor of *Homo heidelbergensis* (500,000 to 230,000 years ago), *Homo neanderthalensis* (300,000 to 30,000 years ago) and *Homo sapiens* (200,000 years ago until the present). If so, it seems likely that one group of *Homo erectus* migrated north to Eurasia and became the ancestors of *Homo heidelbergensis* and *Homo neanderthalensis* while the rest remained in Africa and became the ancestors of *Homo sapiens*.

While it is by no means certain, it seems that the Neanderthals became separated from *Homo erectus*, the ancestors of modern humans (*Homo sapiens*), about 370,000 years ago and spread throughout Eurasia, where they flourished for more than 200,000 years. They left no fossil remains south of Gibraltar and Israel, and some of the latest evidence of their presence is found in Gorham's Cave in Gibraltar. They were heavily built, about five and a half feet tall, much stronger than *Homo sapiens*, and had an even larger cranial capacity. They were better adapted to cold conditions and may even have displaced *Homo sapiens* in parts of the Middle East. Not only do they appear to have coexisted with *Homo sapiens* in Europe for more than ten thousand years, but genetic evidence suggests that some interbreeding may have taken place between these two species of *Homo*.

It is worth pausing at this stage to contemplate what lengthy periods we are still dealing with. When we study human history, the two thousand years of the Christian era seems a long time, for we know so much of what has happened within it. But two thousand years is but an infinitesimal moment in time when compared with the six million years since the hominids separated from our nearest extant primate, the chimpanzee. This means that the period during which the genus *Homo* evolved, and then divided into its various species, is three thousand times the length of the whole Christian era. During that time much genetic change took place. It was halfway through this period that our *Australopithecus* ancestor Lucy appeared, and little less than a further three million years before the advent of *Homo sapiens*, creatures who had the same physical anatomy as ourselves.

Furthermore, only about seventy thousand years ago a supervolcanic eruption, one of the largest that we know about, took place at Mount Toba in Indonesia. It plunged the planet into a dark winter that lasted six to eight years and led to a further thousand years of cold weather. Some have claimed that this drastic climate change resulted in the reduction of the anthropoid population to some ten thousand, or as few as a thousand breeding pairs. If so, human evolution almost came to an end at this point. Others have claimed that most anthropoids came through relatively unscathed. In any case, as our presence here shows, *Homo sapiens* survived. So also did Neanderthals, who did not become extinct until about twenty-four thousand years ago.

The Mount Toba eruption simply serves to underline the somewhat precarious existence *Homo sapiens* had since first appearing in Africa about 200,000 years—that is, one hundred Christian eras—ago. Beginning some 125,000 years ago our species made migrations out of Africa at various times to populate Eurasia and finally the Americas. When the last glacial age lifted about 10,000 years ago, the planet entered what has been called the Goldilocks period, a time when the climate has been neither too hot nor too cold. By that time *Homo sapiens* was the only surviving species of the genus *Homo*. Evidently the size of our brains had given us some survival advantage. Instead of remaining dependent on the very slow evolutionary process of becoming genetically adapted to the prevailing conditions of nature, we employed our increasing skills in the invention and use of tools and in our potential for cooperation to manipulate our environment in such ways that we not only survived but eventually flourished. Indeed, we humans now dominate Earth, having recently attained a global population of seven billion.

Where did we come from? That was the question posed in the introduction. What the story so far tells us is that we came from the earth. However much, during early historical times, our ancestors may have regarded our species as a race apart from all other earthly creatures, or even strangers come to Earth from another world, we can now understand how integral a part we are of the biosphere that has slowly evolved over some four billion years. We are physically related by varying degrees to all the other forms of life on the planet, and if we go far enough back in time we find that we share a common ancestor with every other species that has ever lived.

But the story of human origins is not yet complete. The *Homo sapiens* of forty thousand years ago probably differed little from us in physical anatomy and physiology, but a great gulf still divided them from us. We acknowledge that gulf whenever we describe our behaviour as being either human or inhuman, humane or brutal. The way we experience life today, what we categorise as human nature, had not yet emerged in our ancestors of forty thousand years ago. They still lived an essentially animal existence, not unlike that of gorillas and chimpanzees today. What has brought about the difference between us and the other higher animals is not simply our DNA, but the human culture that we have created and by which we have been shaped. We humans live and work together in highly sophisticated ways. We now take pride in being human, and we speak of human values and human rights. To be human is to be able to think, to feel, to make decisions and to plan our lives. How did we change from being animals to being the humans we are today? What lies at the basis of human culture and civilisation? As we look back forty thousand years, it is only natural to wonder what was missing, what yet needed to come into being. That part of the story of our origins we have yet to explore.

# Logogenesis

## The Coming into Being
## of Language

### *From 2,000,000 Years Ago*

After this long evolutionary process, *Homo sapiens*, our earliest human ancestors, finally emerged—at least in physical form. But until about two hundred thousand years ago, they still lived an animal existence. They shared with all animals the same basic needs—water, food and the sexual urge to reproduce. They shared with all mammals the inherited instincts needed to protect themselves and their young from danger, though to be sure some mammals were more gregarious than others. They shared with the apes nearly all the DNA that determined their anatomy and physiology. Probably, as was the case with all sentient beings, their consciousness still focused on their immediate environment, for it is likely that they had little sense of the passage of time and therefore lived in an eternal now. They would of course have been aware of the changing seasons, for these caused them to develop instinctual patterns of behaviour by which they responded to the yearly cycle. We still observe in awe and wonder the way such instincts impel migratory birds to gather at the appropriate time and travel great distances together.

How, then, did humans come to be so different from all other animals? It was not simply the capacity to think, for all animals and birds are capable of some kind of thought, and however much it may vary among species, they possess some degree of memory. I witnessed these faculties in operation one day when I saw a duck leading her twelve ducklings across a road. Upon

61

reaching safety the duck rapidly surveyed her brood. I doubt if she actually counted her young ones (for even I had a little difficulty doing so as they moved about) but she quickly noticed that one was not there. Hurriedly waddling back across the road, she found the missing duckling and escorted it across. Clearly, that duck displayed evidence of memory, of assessing what she observed and of a simple form of reasoning that prompted and directed her behaviour. This far more than instinctive responses shows the presence of what may be called a mind, though there can be some debate as to the definition and most appropriate use of that term.

It is not simply the capacity to think that makes humans unique, but rather the quality of our thinking. Though it cannot be demonstrated with certainty, there was probably a time when the thinking that went on in the brain of *Homo sapiens* was not significantly different from that of Neanderthals, or even chimpanzees. What we do know is that very slowly and over a very long period human thought advanced, eventually creating the great gulf that now exists between humans and all other animals. What chiefly enabled this to take place was the advent of human language. But it is not the case, as some have argued, that language gave rise to thought. The capacity to think certainly preceded language; however, as language began to evolve it greatly enhanced the depth and precision of thought, enabling humans to become the thinking creature *par excellence*. In attempting to understand what human beings are today, the importance of the emergence of language cannot be overemphasised.

Indeed, language now so engulfs us that we often fail to recognise how dependent we are upon it; living as we do in a world of language, we have come to take it for granted. This has been the case ever since we were little children learning our first words. We tend to treat words as basic elements of the natural world, and often it is not until we encounter another language (notice how we instinctively call it foreign!) that we even begin to realise how dependent we are on our mother tongue, and how isolated we feel in the presence of those who do not share it.

When we begin to study the role and creative nature of language, we learn to appreciate its importance in the evolutionary story. The advent of language further demonstrates how the evolutionary process is punctuated by successive spurts of creativity; after the Big Bang came the galaxies, then planet Earth, then life, then mammals, then the genus *Homo* and then language. The potential for creativity possessed by the universe is seen not only in the evolutionary process as a whole, but also in the way its successive products—fish, birds and animals—actively participate in the creative venture. Birds create nests and spiders create webs. After all, nests do not grow ready-made on the trees; birds construct them, the urge to survive having led them to evolve the instinct to do so. This creative activity is taken a stage further when creatures not only become creators but also invent tools to assist them in the construction process. We have observed in the last chapter how *Homo habilis* was so named because of the tt ability shown by these early ancestors.

The most important tool created by *Homo sapiens* is language. We can speak of it as a tool because (as we shall see in Part Two) it has been the medium through which many other elements of our culture have been created. Language is not one of the gifts bestowed by God, as the early biblical thinkers assumed it to be. Rather, it is a wonderfully creative tool that the human species itself has invented. Moreover, it is far and away the most important tool ever created by any species.

Yet it is not entirely true to say that humans created language; it is rather that human language and the human species evolved in tandem. The gradual but slow development of one nurtured the equally gradual development of the other. Only with the advent of modern humankind are we able to look back and claim that we humans, collectively and over a very long period of time, invented language, but it is equally true that the evolution of language has made us the human creatures that we are. As odd instances have shown, human infants who have been reared by animals do not develop into human adults and at some point fairly early in life even lose their capacity to learn language and become fully human.

Of course language was not the product of any one individual person, but of the evolving human species as a whole. Thus, although no human individual is self-made, there is a sense in which the human species is self-made, or better yet, that humanity is a self-evolving species. The growing powers of imagination and creativity within the human psyche enabled the human species, generation by generation, to create and improve this unique tool that is both inherent and essential to human existence.

Language is basically a medium of communication, but that in itself does not make it unique. Most animal species have developed their own specific modes of communication; bees communicate through dance and headbutting, and some creatures share information through antennae or tentacles. Birds communicate through their calls, and all animals have developed various cries to attract the attention of their own kind and to convey simple bits of information. But do they have language? To be sure, we sometimes speak of animal modes of communication as their 'languages', but this is an equally metaphorical use of the word as that found in such phrases as 'body language'.

Human language is a unique medium of communication. First, as its derivation from the Latin word *lingua* (tongue) shows, it refers to the way we humans, by tongue, lips and palate, manipulate the various sounds we can make as air passes through the larynx. Second, these sounds are much more than meaningful cries, even though they may have originated as such. The reason human language far surpasses the forms of communication developed by all other species is that these sounds are symbols. Except for a relatively few onomatopoeic words, the sound of a word bears no inherent relation to the meaning it conveys. This is illustrated quite simply by the fact that different languages use quite different combinations of sounds to convey the same or similar meaning. In language, vocal sounds have become *symbols of meaning.*

Linguists further analyse vocal sounds into consonants, vowels, syllables, phonemes and morphemes; the shortest freestanding combination of sounds capable of conveying meaning are what we call words. When words are gathered and arranged to convey more complex meaning, such as an idea or a thought, we call that a sentence. Each language has developed not only its own collection of words or vocabulary but also its own syntax—the rules by which words are joined together in sentences to convey thoughts.

Language has become so much a part of the human condition that we cannot remember a time without it. Our earliest memories date from after we began to speak, and we pick up the syntax of our mother tongue well before we are taught the formal rules of grammar in school. After all, throughout most of human history formal schooling has not existed. So how did we humans come to be the speaking apes? How did the various human languages originate? Today the scientific study of human language is known as linguistics, and although it has only very recently made its appearance in our universities, our forbears were engaged for several millennia in philology, the study and acquisition of languages other than their own.

The origin of human language has been described as the most difficult problem in science. Indeed, in 1866, shortly after the publication of Darwin's epoch-making book, the Linguistic Society of Paris imposed a ban on any further discussion of the evolution of language! Nevertheless, a great deal is still being written about the subject, though much of it may be little more than informed speculation.

When did language originate? Whereas anthropologists have learned much about the evolution of our physical bodies from the study of fossils, language leaves no traces of its origin and early development; therefore, linguists have no evidence with which to work, and must depend largely on conjecture.

Fossils have been helpful, however, in determining when human anatomy developed to the point that speech became possible. The advent of human language was not possible before certain developments took place in the larynx and the brain grew to something like its present size. This is why it will never be possible to teach chimpanzees to speak. Although Neanderthals had an even larger brain than *Homo sapiens*, it is uncertain whether they had developed language before becoming extinct about twenty-four thousand years ago. Of critical importance was the lengthening of the larynx. That this development began no more than several hundred thousand years ago is evident from the fact that it does not take place in babies until three months after birth. This delay enables infants to suckle and breathe at the same time, something we adults cannot do. We pay the price of vulnerability to choking for the inestimable advantage of being able to speak.

The fossil evidence shows that the physical features necessary for speech began to appear about two million years ago, but were not fully developed until about two hundred thousand years ago. In addition to the extension of the larynx, the brain needed both to become larger and to develop a special facility now known as *Broca's area*—a feature that chimpanzees do not possess. Then the development of speech began to make its own demands on the brain, for we now know that the left frontal cerebral cortex is essential for symbolic language. Although language could not have developed before the advent of these anatomical changes, the mere possession of them did not automatically lead to the production of language.

*Homo sapiens* probably inherited a simple form of communication from its predecessor, but this medium would have made more use of gestures than of vocalisations. *Homo erectus* seems to have been the first hominid to form small societies that lived and hunted in groups, invented complex tools and cared for its infirm members. To facilitate these activities *Homo erectus* must have developed a relatively complex medium of communication—one we might call a proto-language—that probably consisted of meaningful signs made by hands, arms and facial gestures and accompanied by cries and grunts. Such a means of sharing information could well go back at least a million years, perhaps two million.

Trying to establish a specific date for the origin of language may be of little importance, for as with all evolutionary developments it is probable that the initial stages were characterised by exceedingly slow and gradual progress before a period of acceleration brought it to completion. Certainly the potential for semi-vocalic proto-language was present among the Great Apes when they developed gestures and sounds to communicate with one another some fifteen million years ago, yet the mature development of vocal language probably emerged less than one hundred thousand years ago.

Then how did language subsequently evolve? As we look back from our current vantage point, it seems quite natural to assume that language evolved as it has, simply because of the obvious advantages it gave to those using it. But when we pose the question within the context of an evolutionary process that proceeds by chance events and random mutations, we can see that such an appeal to hindsight may be misleading. Even two hundred thousand years ago it was neither certain nor even probable that the species *Homo sapiens* would create language as we know it today. For like all evolutionary creations, language resulted from a series of fortuitous events, beginning with the physiological changes in the larynx and brain.

Members of all species that we classify as gregarious are motivated by their nature to develop some form of communication. We have already mentioned the cries and gestures that various creatures use to communicate, and we may justifiably assume that our distant ancestors shared these with other creatures. It is almost certain that vocal language in humans first evolved as a *supplement* to gestural forms of communication. Indeed, even now most of us instinctively resort to gesture, both when we cannot find the right words to express a thought and when we wish to dramatise or emphasise the words we use. And when someone asks for directions, we nearly always respond by pointing, to supplement our verbal reply. Michael Corballis, one of the supporters of this gestural theory of language origins, humorously tells how he struggled to persuade a prominent linguist of its validity, only to have his antagonist summarily dismiss it with an eloquent wave of the hand!

It is remarkable how people, however articulate, will unconsciously resort to hand movements when the right words do not come to mind. Note further that we commonly speak of the sum total of a person's physical movements, intended or involuntary, as "body language". And we still resort to gestures when we wish to convey our feelings: we grimace; we shrug our shoulders; we show pleasure or sorrow by facial expressions.

Now it is generally true that the more successful any activity proves to be, the more we are motivated to refine and improve it. And even in its initial stages, vocal communication was quite an advance on gestures if only because it freed the hands to perform other functions. Likely enough, this is why it was used early on to supplement gestural communication, even though this order has since become reversed. In short, it seems all but certain that grunts, cries and babble similar to those of human infants today, and those still used by other members of the ape family, were the beginnings of vocal language.

We may still gain insight into the evolution of vocal language by observing language development in children—though we must allow for the fact that a child's progress in less than two years may be the equivalent of twenty to fifty thousand years for our ancestors. Babbling infants, though already genetically pro-grammed for language in a way our distant ancestors were not, nevertheless suggest—or perhaps even recapitulate—the long experimental and developmental stages that language passed through.

It seems probable that the very first words to be created were the names of visible objects—things that could be pointed to (notice the early relation of language to gesture!). Indeed, that is the way we introduce children to the world of language now. One of the most common questions a child learning to speak will ask is simply, "What is that?" And among a fond parent's favou-rite prompts is, "Where's the doggie?", or some such question asking the child to point to the object represented by the naming word. Giving names to objects not only enhances the commu-nication process, it also helps us to create an ordered world, one that thereby becomes known to us and hence safe. That which has not yet been named seems more threatening by virtue of being unknown. Even we adults usually feel relieved when our doctors can give a name to our malady, even if they cannot sug-gest a cure. In like manner, naming the objects they encountered enabled early humans to gain a degree of psychological mastery over their situation. After that came words like *sit, run* and *hide* that referred to common actions. This stage of vocabulary cre-ation may have lasted for a very long time before the invention of syntax allowed for further development.

The first step in the evolution of syntax was to make simple statements by relating words without the use of connectives: for example, "Storm bad" or "Jack strong" or "Bird fly". In all likelihood this is how primitive grammar began to take shape. It can be observed in the way children develop language skills, and some languages still retain the residue of this simple association of objects. But once words were put together to make state-ments (or what we now call sentences), language began to evolve much more rapidly. Eventually each language developed the complex structure that we call its syntax. The speed and apparent ease with which young children learn the syntax of their mother tongue simply by listening to it, without ever being specifically taught it, suggests that we humans have been using the tool of language for so long that we now inherit some sort of capacity for language.

But if language evolved to be such an efficient medium of communication, why do we not all speak the same language? That question so puzzled even the ancient biblical writers (though they were aware of only a few languages) that they created a story to explain it. (As we shall see in Part Two, storytelling was the primitive method of providing explanations for the basic questions asked by inquiring minds). Believing all humans to have descended from our mythical first parents, Adam and Eve, it seemed self-evident to them that, as they said, "At first, the people of the whole world had only one language and used the same words". So they composed a story based on the building of the ancient ziggurat at Babylon: This immense tower, roughly the size of an Egyptian pyramid, was constructed to provide people with a place of safety in times when the Euphrates flooded, but the biblical writers portrayed it as an attempt to reach heaven and challenge the power of God. In response to this intolerable display of hubris, God put an immediate stop to their project by replacing their common tongue with such a host of different languages that they were thrown into confusion, prevented from communicating and thus unable to collaborate on any future projects.

Of course this ancient myth no longer provides a valid explanation, and we still face the question of why so many languages developed and whether a single proto-language ever existed. Linguists have been surprised by the sheer number and enormous diversity of languages discovered in pre-literate societies. The population of New Guinea and surrounding islands, for example, numbers only ten million, but they speak about

1,150 languages! The islands of Vanuatu have a population of less than two hundred thousand, yet they speak 105 different languages. On the other hand, no human population has been found without language—and this despite the fact that some, like the now-extinct Tasmanian aboriginals, were likely separated from the Old World continents as long as forty thousand years ago.

How did a species with a common origin come to be speaking as many as 6,500 languages? First, linguists observed that languages can be readily classed into families. The so-called Romance languages—Italian, French, Spanish, Portuguese, Romanian, Catalan, Sardinian—all clearly evolved from the Latin of ancient Rome as it spread during the Roman Empire. Latin and Greek, in turn, belong to the Indo-European family of languages that includes the Celtic, Germanic and Slavonic groups, as well as Farsi and Sanskrit. What we learn from the evolution of families of languages is that language diversifies over time in much the same way as a single animal species diversifies into a genus of several species.

Thus, language displays the same characteristics as life itself. Languages are born and languages die. Indeed, we commonly speak of 'living languages' and 'dead languages': Cornish has died out, Gaelic has barely survived, and Welsh is very much alive in North Wales. Thirty years ago it was widely feared that Maori was dying, and strenuous efforts are now being made to revive it. Languages may even be resurrected, as in the case of Hebrew, which not long ago was found only in academia and the synagogue, but is now the living language that unites the people of the new state of Israel.

Linguists have classified all known languages into about seven families. Because of the links that still join them, each family can probably be traced back to a proto-language. Linguists further believe that the Indo-European languages can be traced back to a supposed Proto-Indo-European language spoken by a particular

tribe about six thousand years ago. However, since this classifica-
tion must be seen as provisional and the relationships between
the families are still uncertain, it is not at all clear whether all
known languages can be traced back to a single original.

Some linguists embrace the multi-regional hypothesis, according to which different language families evolved independently
on different continents. Such a view seems very plausible when
we consider how different from one another some of the families
are—compare the Indo-European languages with Chinese and
with some of the African languages, for example. Yet a majority
of linguists favour the hypothesis of *monogenesis*, having concluded that all human vocal languages have descended from a
single proto-language that has been given the name Nostratic.

Since the 'Out of Africa' hypothesis maintains that all humans
alive today are descended from a woman (Mitochondrial Eve)
thought to have lived in Africa some 150,000 years ago, then
this hypothetical proto-language, Nostratic, may date from that
period. Such an expansive time period allows ample opportunity
for the evolution of the many languages of today, particularly
when we observe how quickly a language can mutate. Even
among those of us whose mother tongue is English, few can
readily understand the Chaucerian English of only six hundred
years ago, and most experience some difficulty with the merely
400-year-old language of Shakespeare.

Present evidence leads us to suppose that the evolution of
language had not proceeded very far before the gradual spread
of the human species around the globe meant that diversification began to set in. Over more than one hundred thousand
years humans moved out of Africa to all continents and most
habitable islands. As groups lost contact with one another, the
supposed proto-language—a living and changing entity, after
all—spawned a number of quite different forms, much as geographical dialects quickly emerge today. Each new language developed a life of its own, its success depending on the fortunes

of those who spoke it. Thus the evolution of languages parallels the evolution of species. Within historical times languages have been spread by sudden ethnic expansions and invasions. The swift rise of Islam had the effect of greatly extending the area where Arabic was spoken, and European colonialism spread Spanish, Portuguese, English, Dutch, French and German around the globe.

If all the known languages could be traced back to one proto-language, we would expect to find some simple deep structure that remains common to them all. For example, some think that it was through the evolution of language that humans developed their sense of time, and this suggests that languages lacking verbal tenses reflect an early stage in language development. Of particular interest in this respect is the experience of a Christian missionary, David Everett, who went to a remote tribe in Brazil known as the Pirahã to convert them to Christianity in 1977. During the six years it took him to master their language, he made the intriguing discovery that they had little sense of time and no understanding of past or future. They lived in the present. Their language had no tenses and their culture had no myths or stories of the past.

That fact both points to and underlines an extremely important consequence of the advent of language: human language originated as a medium of communication, enabling people to improve their skills and pass their knowledge on to the next generation. It enabled the already gregarious human species to develop into more closely-knit communities. This in turn improved their chances of survival, but soon began to do a great deal more. Whereas language originated to make communication possible, it became the medium through which a new kind of creativity began to take place. This unforeseen and unintended breakthrough came with the telling of stories, what the Greeks called *myths*; their word *mythos* literally means 'something told by word of mouth'.

Modern people tend to view themselves as sophisticated and hence dismiss the myths of ancient peoples and isolated tribal communities as primitive or foolish. What they fail to appreciate is that these creations represent a very important stage in the way humans came to understand and respond to the world. Stories were invented to explain natural phenomena. They were a primitive form of knowledge or 'science'. Stories led to the emergence of abstract or pure thought, and this is the chief reason why the advent of language must be judged a critical transition point in the long story of cosmic evolution. The advent of language has led to the creation of a whole new kind of world—the *world of human thought*, whether imaginative, rational, or irrational. And this world, though non-physical, is just as real as the physical world whose story we have been tracing so far. But to say more than this is to trespass on the topic of the next part of this book. So we now turn to the story of how this new world evolved.

# The Evolution of the Human Thought World

# CHAPTER 6

# Noogenesis

## The Coming into Being
## of the Noosphere

### From 50,000 Years Ago

Even though this book continues chronologically, we now move from Part One to Part Two in order to emphasise the significant transition that was brought about by the advent of language. Part One told the story of physical evolution, concluding with the arrival of an ape that speaks. This speaking ape, *Homo sapiens*, has now come to dominate Earth and lord it over all other creatures. Even more important, the advent of language has enabled this speaking ape to ask questions and search for answers. Not only can this species speak its mind, but its members are able to contemplate the universe in all its wonder.

If we reflect on this emergence of human self-consciousness within the context of the long story of evolution, we should find it as awe-inspiring and groundbreaking as the emergence of life itself some four billion years earlier. It is truly mind-blowing to realise that the self-evolving universe, in spite of taking eons to achieve it, has brought forth a creature through whom it can now look at itself and ask questions about how it all began. The time when the cosmos first thought about itself through us was surely a moment of emerging self-consciousness *par excellence*!

In Part Two we are, to borrow the slogan over the entrance to Harrods massive departmental store in London, "entering another world"—one that may be called the world of human thought. Because language enabled humans to share their

thoughts and eventually to bounce their thoughts off one another, human thought began to create a new kind of world.
Indeed, this new world of thought began to expand outwards
rather like the way the physical universe expanded after the Big
Bang. And even though it is not physical, this world of thought
is just as real as the physical universe. Thus, the advent of language may be said to have initiated another kind of evolution—
the evolution of the human thought world.

It is important to acknowledge (as we have above) that
thought precedes language. This becomes obvious every time
we confess that we cannot find the words to say what we are
thinking. Because experimental psychologists have shown that
hominids are capable of quite complex thoughts, it is clear that
thoughts germinate in the human psyche and grow to some
maturity before they are expressed verbally—and many of them
never do. When thoughts find expression in language, one of the
functions of the brain is to provide a word-finding facility, and
this function diminishes in quickness and proficiency as we grow
old. What the advent of language did was provide a tool that enabled humans to think more clearly and to attain a much higher
and more abstract level of conceptualisation. This is illustrated
by the quip about the professor who, on being asked what he
thought about a particular topic, responded, "How do I know
what I think until I hear what I say?"

We must pause here for a moment to take cognisance of a
little-realised truth that now becomes strikingly obvious: The
story of the evolving universe related in Part One could not have
been told without words. Indeed, no story of any kind could be
told before the advent of language. This realisation places us in a
strange and paradoxical situation. The story in Part One properly
belongs to the world of human thought, the account of which
we are only now beginning. And when we tell a story, we usually
do so from a standpoint outside of the places, objects or people
we are talking about. But however much we may try to hold a

picture of it in our imaginations, the universe has no outside. We ourselves are an integral part of the universe we are trying to understand. Neither is there any 'outside' to the world of human thought. We were initiated into it when we learned to speak and we have lived within it ever since. This world of thought, transmitted to us within and by our culture, has played a large role in making us who we are. Therefore the story of the universe must be told not only from within the universe but from within the world of human thought.

To make matters even more complicated, what I have loosely called "the world of human thought" is not characterised by the unity that the term implies. It is not so much *one* world as a *multiplicity* of worlds, each belonging to the language and culture in which it is used. Although modern globalisation is beginning to draw this plurality of cultural worlds into one, it still has a long way to go if it is ever to achieve such a result. This means that every attempt to tell the story of the universe will display a cultural bias, and try as we may, we cannot escape it. As yet, the definitive or undisputable story of the universe cannot be told. Indeed, it will be clear to most readers that this story is at present being told from within the world of Western culture.

All of this forces us to recognise a clear distinction between the universe itself and the story of its coming into being. Though it may seem trite and altogether unnecessary to say so, the story of the universe is not at all the same thing as the universe itself. The *story* of the universe is the universe *understood and interpreted* by human minds. An interpreted object is always at least one step removed from what the words point to and describe, an insight that the philosopher Immanuel Kant (1724–1804) called to our attention when he made a clear distinction between our perception of a thing and the "thing in itself".

Let me explain further. The data assembled in the previous chapters are facts and thoughts that have entered enquiring human minds and been expressed in human language. But however much we may like to appeal to such a claim, "*bare* facts" do not exist: all facts, ideas and statements are expressed in words of human languages, and by virtue of being the products of human minds they become *interpreted* facts. Thus we must be careful to distinguish between the physical universe and our verbal description of it.

This distinction has been strikingly expressed in *Inventing Reality,* a book by physicist Bruce Gregory:

> There is a sense in which no one, including philosophers, doubts the existence of a real objective world. . . . [But] the minute we begin to talk about this world . . . it somehow becomes transformed into another world, an interpreted world, a world delimited by language

In other words, the world we *know* with our minds, and *talk about* with our lips, is intrinsically different from the reality we infer from sensory impressions because it is a world we have already interpreted. Our world, the world to which we respond in the way we live, is not reality itself but reality understood and interpreted through the grid of language and the shared general knowledge we receive from our culture.

Science justifiably prides itself on its objectivity, but it is not possible to study the whole of reality objectively. To make this point, the internationally famous philosopher of science Karl Popper (1902–94) proposed a 'three-worlds' model by which to understand the whole of the reality in which we humans live. The world of human thought corresponds to World 3 in Popper's model.

What Popper called World 1 is the presumed physical universe whose coming into being has been sketched in Part One. It includes all physical objects and processes, not only on this planet but throughout the universe as a whole. It encompasses both organic and inorganic substances. All living organisms are part of World 1—and that includes the human species and the human brain with which we think.

What Popper referred to as World 2 is the world of subjective experience. It consists of all the states of consciousness found among living creatures, from those present in the simplest forms of life right up to human self-consciousness. To be sure, consciousness is difficult to define because it cannot be examined objectively. Each of us knows our own consciousness subjectively, but the consciousness of all other creatures, including other humans, is hidden from us. Consciousness has been described as subjectivity, awareness, wakefulness and the ability to experience or feel. Today we even speak of abnormal or artificial states of consciousness, such as those induced by some religious practices or psychedelic drugs.

Since consciousness is a non-physical reality, it is not surprising that we tend to think of it as something separate from the body. This tendency led the philosopher Plato to speak of the reality of consciousness as the soul, an entity quite distinct from the body in which it temporarily resides, and one that not only survives the death of the body but also existed before it. Plato's influence led to the dualist view of reality that came to dominate Western culture, a characteristic we see expressed in such pairs of opposing terms as body and soul, material and spiritual, natural and supernatural. Even the more rationalist thinker Descartes, who believed that his capacity to think provided clear evidence of the reality of his soul, must be held responsible for perpetuating this dualism of body and soul into modern times.

However, during the twentieth century the recognition of the psychosomatic nature of the human condition led to the widespread demise of dualism, which has largely been replaced by some variety of what is called *functionalism*, in which consciousness is not regarded as an entity with an independent existence, but is recognised as a function of the brain. Thus, when the brain dies it can no longer function, and consciousness ceases.

Consciousness can be described as the state of being aware of one's surroundings. Having reached the age where I find myself waking several times in the course of a day, I frequently observe the stages through which I pass in returning to a state of consciousness. For example, the sounds I hear on waking may alarm me because of their strangeness or they may reassure because of their familiarity. Then when I open my eyes, the amount of light I perceive may indicate whether it is night, morning, or afternoon. Seeing familiar objects shows me where I am, and they may be particularly reassuring if I am waking from a bizarre dream. Likewise, the sight of strange surroundings can be alarming to someone waking up in a recovery room after surgery. When at last I recognise exactly where I am, determine the time of day and am ready to respond appropriately to my environment, I have become, as we say, fully conscious.

But then there arises the issue of the focal point of consciousness. It is possible to concentrate so exclusively on one object or train of thought that we remain quite unaware of some unusual event occurring within the range of vision. Even more important, if our minds become flooded with all sorts of memories and musings, we may become so focused on our inner world that we ignore our immediate environment and respond to it only by what has been aptly termed our 'automatic pilot'. This correlation between absorption with an active inner world of thought and loss of attention to the outside world is well exemplified by the proverbial 'absent-minded professor'.

Because human consciousness is such a subjective and elusive phenomenon, it is not surprising that scientists long avoided subjecting it to empirical study. Indeed, the behaviourist school of psychology, founded by John B Watson and further expounded by B F Skinner in the 1940s, refused to acknowledge that consciousness is a fit subject for scientific examination. More recently, psychologists and neuroscientists have done a large amount of experimental work on consciousness in general and human consciousness in particular. We now know a great deal more about the nature and *modus operandi* of the human brain, and can even connect features of consciousness with particular areas of, and neural circuits in, the brain.

All sentient creatures appear to experience some degree of consciousness, for they demonstrate it by their response to the sensory signals they receive from the environment. Like us, the higher animals all alternate between sleep and consciousness, and perhaps they even dream. That they have memory of objects and places is also evident from their ready recognition of those they have experienced in the past. But consciousness is so private that we have no way of knowing how they experience consciousness. This is true even of our pets, for although we often project our own form of consciousness onto them, we are usually aware that we are doing so. Still, there is good reason to conclude that the consciousness experienced by other primates has some similarities to our own, though it operates at a much less complex level.

Human consciousness is dependent on the many varied sensory impressions that we receive from our surroundings and that our brain deftly collates and manipulates. But humans have the potential to reach a level of consciousness that non-humans cannot, and this is because language has enabled consciousness (Popper's World 2) to construct what Popper called World 3—which he described as the "products of the human mind". To be sure, these products are very real, but like consciousness they are also non-physical. World 3 may be defined as the sum total of human thought and knowledge.

From the time we learn to speak, as we imbibe the culture into which we are born, our consciousness is being enriched and deepened by the impact of World 3 upon us. As we grow and mature we become increasingly aware that we possess an inner world (or mind), whereas animals appear to have no awareness of such an inner world.

Just as it was the possession of a World 3 that gave identity to each of the early cultures, so it is that by being shaped by a

World 3 that we today learn how to live as human beings—by responding to the physical environment, relating to others and eventually establishing our own individual identity. It is the evolution of World 3 that caused the great gulf to open up between human consciousness and that of the other higher animals. Even though much of World 3 is nowadays expressed in physical forms such as books, it is intrinsically non-physical; after all, it came into being long before the invention of writing. It existed for centuries in human memories, and through the medium of language was handed on to all those who shared that culture. This may be illustrated by the fact that the whole of the Qur'an originated in the mind of Muhammad and, before being committed to writing, existed for some years only in the memories of his closest followers, who were appropriately called the Qur'an bearers.

Each human culture evolved its own World 3 in tandem with the language that formed its base. Beginning with names, descriptive terms and concepts, the World 3 of each culture began to expand, eventually to contain stories, poetry, explanatory theories (both true and false), patterns of behaviour, codes of ethics and social institutions. As cognition and human reflection became more abstract and language came to be used more and more symbolically, World 3 expanded further to give birth to religion, philosophy and eventually modern science.

Each of us becomes immersed in a World 3 from the time we learn to speak and listen to what we are being told. World 3 becomes the lens through which we observe and understand the World 1 of which we are a part. This is what we superimpose on the physical cosmos we experience through our senses. It is World 3 that enables us to successfully respond to the impact of World 1 upon us, even though it also keeps us one step removed from the 'thing in itself'. Because we have experienced a World 2, and have lived in a World 3 from the time of our earliest memories, we are inclined not only to take both Worlds for granted but to assume they have been there from the beginning. Because language was a necessary prerequisite for consciousness to produce a World 3, it appeared to our ancestors to be so basic that they could not imagine the world without it. When the biblical thinkers composed their story of the creation of the universe (now found in the opening chapter of Genesis) they simply assumed that language had existed from the beginning of time—and indeed, even before that! They went so far as to declare it to be the creative power that brought Earth and sky into being. As they put it, God the Creator had only to say the words, "Let there be *x*" and *x* came into being. (We speak of this theory of origins as *creation by divine fiat*.)

The author of the Fourth Gospel, following the pattern established in Genesis, went even further. Notice how he began: "In the beginning was the word and the word was *with* God and the word *was* God" (emphasis mine). The word that he used, *logos*, meant something rather more than a word of language. Rather it signified 'reasoned thought', as its derivative 'logical' makes clear. In fact, we could translate the first part of the passage to read, "In the beginning was reasoned thought", with the implication that logical thinking was the 'God' who created the world.

In this way these ancient biblical thinkers stumbled upon an important truth when they declared that everything was created by language. *A* world was in fact created by language, yet it was not the physical universe (World 1), as they assumed, but rather the world of human thought, Popper's World 3. As noted above, it is to our World 3 that the story of the universe belongs. It has been evolving there for quite some time but with increasing speed in modern times.

It is salutary to remember, however, that Popper's three-worlds concept is simply a model constructed by humans to help our understanding. Just as all theories have their weaknesses, so in Popper's model it is sometimes difficult to distinguish clearly between Worlds 2 and 3. This is because from a very early age the visual, auditory, tactile and olfactory signals that we receive through our sense organs are shaped and interpreted by the cultural impact of the World 3 that surrounds us and enables us to become fully human. In consciousness, that is, we are not *merely* conscious; rather, we are *conscious of our environment*—which as we noted earlier consists of objects that are recognisable, known to us because of past experience, and meaningful to us because of how we have learned to interpret them.

Thus, useful though Popper's model can be, it might be helpful to compare it with another concept. Some decades before Popper, Pierre Teilhard de Chardin coined a new term, *noosphere*, for the world of human thought, what Popper referred to as Worlds 2 and 3. Teilhard de Chardin invented the term by combining the Greek word *nous* (meaning 'mind' or 'intelligence') with *sphere*. Just as Earth had become enveloped by the hydrosphere, the atmosphere and the biosphere in turn, Teilhard saw the evolution of thought within the human species as a significant new activity encompassing Earth, which he said now possessed a "thinking envelope". Metaphorically speaking, we could even think of the noosphere as an embryonic "mind of the Earth".

Teilhard was at pains to assert that in the long history of the cosmos the emergence of the noosphere marked as great a breakthrough as the emergence of the biosphere some three billon years earlier. Teilhard saw this as a further and highly distinctive example of his law of complexity/consciousness. He argued that the basic energy of which the universe is composed had the potential for consciousness from the very beginning, but that potential could become a reality only when physical energy had become organised into a sufficiently complex design. The human organism, with its nervous system and brain, was just such a complex design. This, he proposed, is why the evolutionary process took so many eons to produce self-conscious, thinking human beings.

As life evolved—from the unicellular to the multicellular, from the multicellular to the organism, from the organism to the animal, from the animal to the Great Apes, from the Great Apes to the humans—sentient creatures were becoming increasingly complex, particularly in the area of brains and nervous systems. Then came the radical advance in the species *Homo sapiens* when the advent of language introduced a new level of complexity by facilitating interaction among human minds. This interaction not only led to greater social cohesion but lifted human consciousness to a yet higher level, and thereby introduced a new kind of activity in the cosmic process—the noosphere.

The noosphere opened a great gulf between animal consciousness and human consciousness. Whereas the other higher animals think and may even be said to know, it is *only humans who know that they know*. That is the key to understanding Teilhard's notion of the noosphere. Unlike Popper's model, however, this concept does not even attempt to distinguish between mental activity (Popper's World 2) and the products of mental activity (Popper's World 3). All sentient creatures that display consciousness share at least a minimal capacity to think. But lacking language, non-humans remain limited as to how much they can share their thoughts. The advent of language enabled humans to share their thinking and construct a body of knowledge (World 3) that could be transmitted to the next generation. This not only led to the expansion of that world but enabled humans to reach a still higher level of thought. Michael Corballis has labelled this the *recursive mind*, a term that offers a more refined understanding of Teilhard's noosphere, but since Soviet scientists of the time were quick to acknowledge the usefulness of Teilhard's term and many others have followed suit, we shall now adopt it in this book, although we shall also often have occasion to refer to Popper's Worlds 2 and 3.

Just as the biosphere introduced a new kind of cosmic activity on Earth (the emergence of organic chemistry, for example), so the noosphere added another new activity, for it operates within the minds of the human species and is transmitted by language. Just as the biosphere operates according to the natural laws of physics and chemistry, so the noosphere is dependent upon the chemical and biological activities that take place in the brain and nervous system. This relationship between the noosphere and its physical place of origin thus raises a whole host of issues—philosophical, psychological and neurological—that we are still trying to understand. Nevertheless, it is now clear that if at some future time the human species should become extinct, the noosphere would effectively cease to exist, and the only evidence of its existence would be such human artefacts as libraries, waiting to be discovered by some future visitor from outer space.

The evolution of the noosphere may be called *noogenesis*. Though no doubt having undergone a very protracted early development (like most new phenomena in cosmogenesis), it evolved much faster than the human species, dependent as that process has been on small, incremental changes in DNA. Biological evolution is so slow that the anatomy and physiology of the human body has probably changed very little during the last fifty thousand years. In some geographical areas humans have increased in size and in a few they have decreased to pygmy size. Scattered around the globe, we humans have so diversified in skin colour as to become popularly known as brown, black, white, red and yellow. Over time, a considerable loss of body hair impelled our ancestors to clothe themselves. By and large, however, we are physically much the same as the earliest examples of *Homo sapiens*. Yet that same period has witnessed the gradual evolution of the noosphere and the consequent widening of the enormous gulf that now separates contemporary civilised humans from our cave-dwelling ancestors.

Today the noosphere is expanding at an accelerating rate due to what is sometimes called the 'knowledge explosion'. This results in part from the burgeoning of the scientific enterprise and in part from the new media of radio, television and Internet by which it is being transmitted. The noosphere is expanding so fast that we cannot help but be aware of its growth; only during the latter part of the twentieth century did we begin to speak of the 'generation gap', and now we grandparents find ourselves being taught by our grandchildren how to communicate electronically!

The noosphere is both a human activity in which we participate (World 2) and a body of human knowledge with which we have a symbiotic relationship (World 3). It continually shapes us and we continue to construct and reshape it. Throughout the rest of this book we shall focus exclusively on the evolution of the human species, and especially on how it has contributed to its own evolution by constructing and shaping the noosphere.

But how did the noosphere come into being? Even if we go back to the beginning of recorded history, say five thousand years ago, we find we are dealing with people much like ourselves. Further back than that we must depend on conjecture, and we need to remember that if we attempt to retrogress from the present period of rapid cultural change into the past, we must acknowledge an ever-decelerating rate of development the further back we go. With those provisos in mind, the story might go something like this:

The noosphere emerged very, very slowly from an imperceptible beginning during the time when the advent of language first enabled small groups of humans to share their thoughts more successfully and with greater precision than they had previously done by means of gesture. The thoughts they communicated were very simple at first and not at all abstract, since they dealt with the practical affairs of day-to-day life, and although these basic thoughts were to become the seeds of future human cultures, it would be a very long time before they evolved into those primitive cultures that have left some traces. Just as their proto-language mutated and diversified into a host of languages as the human species spread around the globe, so the language-based noosphere did likewise in its earliest stages. Each ethnic group developed its own language and slowly assembled a body of cultural knowledge that it handed on to the next generation. Each community learned by experience how to respond most successfully to its environment and how to use its imagination to invent stories and retell stories from the past. Each ethnic group evolved its own account of where it came from. Thus, their particular version of the noosphere provided them with their cultural identity. In these pre-literate societies the content of the noosphere (World 3) remained where it originated—in the minds and memories of people. It was to be a long time before it could be transcribed and preserved in scrolls and books.

We shall now attempt to sketch the development of the content of the noosphere, variously referred to above as the world of human thought or World 3. This sketch will be done in three successive stages: The first we could call the Age of Stories; the second, the Age of Speculative Thought; and the third, the Age of Empirical Knowledge. For reasons that will no doubt soon become clear, I have entitled the first stage *polytheogenesis*—the coming into being of the gods.

# CHAPTER 7

# Polytheogenesis

## The Emergence
## of the Gods

*From 20,000 Years Ago*

As the human species slowly spread around the globe from its place of origin in Africa, it also kept dividing into communities. It is thought that the optimum number for a viable community in those times was about 150 individuals. They were hunters and gatherers whose survival depended on what the natural world could supply. No doubt some of these communities perished when excessive cold or drought led to starvation, but those that survived became the forerunners of the later ethnic groups. As those early human communities spread into new territories and lost touch with one another, both language and culture diversified. Each of them slowly accumulated and evolved a particular body of cultural knowledge that was passed on from generation to generation and provided the group with its ethnic identity. It is this body of cultural knowledge that may be called their *thought world*, their unique version of Popper's World 3.

Since well over six thousand languages survive to this day, we may assume that an even larger number of thought worlds must have evolved. The sum total of them all, disparate as they were, constituted the embryonic noosphere. Thus, in its initial stages, the noosphere was by no means a single phenomenon but was composed of a large number of independent entities that were embedded in human minds, shared among members of the same group and transmitted to the next generation. These came to differ more and more from one another as the human species dispersed around the globe and as languages diversified. The elements they had in common were due to the similarity of their experiences and from the human condition they shared. And just as we find family groups among the languages in which these cultures were transmitted, so we can also recognise families of thought worlds that evolved from an earlier common origin.

Of course, we have no way of learning the content of these primitive human thought worlds, for at the very least some fifty thousand years were to pass before the creation of the first extant evidence of what early humans thought. And while the slow development of language must have been a key contributor to the evolution of the human mind, it is likely that a long time elapsed before human thought began to differ in any significant way from what went on in the minds of other hominids or even those of the other Great Apes.

The following attempt to sketch what may have slowly developed in primitive human minds is based on three sources of information: First, we have the extant cultures that have been isolated from the rest of the world for up to forty thousand years, such as the Australian aborigines and the tribal cultures of New Guinea. Second, we now have a fairly extensive knowledge of ancient cultures in Mesopotamia, India and China from some six thousand years ago. Third, we can compare what we know of these with the way our personal thought worlds evolve out of nothing from infancy onwards. Our individual thought worlds evolve exceedingly quickly because they are largely shaped by the already-existing culture into which we are born, so that what takes us only a few years was spread over a hundred thousand years for our distant ancestors.

The first conclusion we may draw from a study of these three sources of information is that the primitive thought world probably initially consisted of the names of the objects, living creatures and phenomena that were frequently encountered in their immediate environment. As we earlier noted when discussing the evolution of language, names construct a world that is 'known' and psychologically possessed. As children learn language they not only repeat what they hear, but they also keep asking, "What is that?" Names act like signposts used to mark out new territory. As members of a primitive community learned from their forbears how to name common objects and experiences, they were at the same time coming to feel 'at home' in the world they shared with their kin.

The second conclusion we may draw is that the content of the primitive thought world expanded because of the innate curiosity that is typical of both humans and most of the higher animals ('curiosity killed the cat'). Just as two-year-olds are persistent questioners—Why did that happen? How does that work? Where did I come from?—so these and similar questions kept arising in the slowly evolving minds of early humankind. But whereas children today receive answers from their parents, our primitive ancestors took a long time to create a collection of answers.

The third conclusion we find is that they did this by drawing on their imaginations to create stories. It is fascinating to observe that all three of our sources of information employ the same dominant method of passing on knowledge about the world: by telling a story. Little children have no sooner begun to learn their mother tongue and to construct short sentences than they become fascinated with stories and want to hear more. And collections of stories are exactly what we find at the heart of the body of knowledge being handed down orally in tribal cultures and being committed to writing in the ancient centres of civilisation.

In all three areas the stories have been crafted by the creative imagination of human minds; they are imaginary rather than factual or historical. In this respect the mind of a young child, the tribal mind and the ancient mind are somewhat alike in that what captures their interest is the story itself and not when or where it takes place. Furthermore, they generally do not distinguish between fiction and history, for it is only in recent times that we have become concerned with making such a distinction. Even to this day, as a glance around the average public library quickly shows, people spend more time reading fictional than factual accounts. In the course of human history billions of stories must have been told, but in this book we can mention only a few examples, and these are cited in order to illustrate specific points related to noogenesis.

In the stories found in contemporary tribal cultures and in the ancient world, along with those still enjoyed by children, no clear line is drawn between the imagined world and the real world. For in the minds of children, ancients and tribal people, all things are deemed to be possible. The culture of the Australian aborigine includes a concept translated as 'Dreamtime' that expresses rather well how tribal people probably understood the mysterious world that surrounded and penetrated all that they could see and touch. In our own dreams the most extraordinary things take place, often in quite illogical sequence, for in dreams even linear time seems to lose its normal significance.

As depth psychologists have recently discovered, our dreams occur in our unconscious mind, a segment of the human psyche hidden from conscious awareness. Embedded in our unconscious are the memories and impressions of all our past experiences, particularly the more traumatic ones. These are the raw materials that motivate and help to shape our dreams and emerging thoughts. We frequently observe that in dreams our psyche can create objects, creatures and situations that we have never observed or experienced. Further, what we experience in our dreams can seem so real that we can wake from a nightmare in a state of terror. In other cases the events we experience while dreaming are so normal that afterwards we have some trouble deciding whether they really happened or not. Only recently, for example, I had a dream within a dream from which I had seemingly awakened and found myself explaining my dream to others, only to find myself waking up still further! If this can happen with us, how much more often and more vividly must it have been for our very early ancestors, for whom distinguishing between dreams and reality was considerably more difficult.

Just as in dreams we encounter objects and situations we have never experienced, so in our waking hours we can imagine objects and conceive ideas that we have never thought of before. The human psyche is amazingly creative, and as the seedbed in which all our thoughts germinate, the unconscious can generate mental images of objects and creatures we have never seen with our eyes. That is why in the folklore of even the quite recent past our forbears talked about fire-breathing dragons, hobgoblins, fairies, angels and all manner of life forms that are now dismissed as imaginary.

This is but a tiny sample of the visual images that human minds have created by reflecting on and trying to understand the extraordinary and mysterious world in which people found themselves. No doubt the earliest concerns of our primitive ancestors were about the basic needs they shared with other animals, needs that had long shaped the instinctive behaviours that promoted the preservation and regeneration of the species. They needed the food and drink that their environment provided; they needed protection from wild animals, storms, floods and extreme cold and heat. In dealing with these vital issues they could not help but become aware of the changes that kept occurring in their environment, some of them more or less regularly and some of them very unexpectedly. Quite naturally, then, these were the topics that their creative unconscious worked with.

The world around them was so full of movement, activity and change that it seemed to be alive. As H and H A Frankfort, scholars of the ancient Middle East, made clear in their book *The Intellectual Adventure of Ancient Man*, the primitive mind did not at first distinguish between animate and inanimate objects as we do. Primitive humankind was not aware of an inanimate world but rather viewed everything as potentially alive—and therefore treated everything in a personal way. Little children still do; a two-year-old bumps her head on the corner of the table as she runs by and immediately turns to the offending object and cries, "You naughty table!" In much the same way the ancients projected their own subjective experiences onto the world around them.

They did not ask *what* caused a storm but *who* caused it. To them it seemed self-evident that natural events like storms, spring growth and earthquakes were caused by a personal will similar to their own. Just as people today are often quick to explain an unusual event as an 'act of God', so the ancients attributed such happenings to an unseen personal being who either resided in the storm or caused it. Thus, the ancients conceived of the world as a place inhabited by a race of invisible beings or spirits. This is how the gods came into being.

In the earliest stages of the process, the whole external world was thought to be permeated by some invisible and intangible power. We find the residue of this notion in terms like *mana* and *spirit*. Mana is a term by which Melanesian and Polynesian people refer to a hidden force that may reside not only in people and animals, but even in what we regard as inanimate objects.

In Western culture we encounter the residue of this ancient concept chiefly in the word *spirit*, which is still commonly used and difficult to define. It is clear that the concept developed out of the experience of both wind and breath, for in many ancient languages the same word is used to mean 'wind', 'breath', and 'spirit'. The air that surrounds us and on which we depend was once a great mystery. Indeed, it was not until the advent of modern science that humans at last realised that the atmosphere (being composed of gases) is just as physical as solids and liquids.

The mystery about spirit (or wind or breath) was due to it being invisible, powerful and unpredictable. This is clearly reflected in the words that the writer of the Fourth Gospel placed on the lips of Jesus Christ: "The spirit (wind) blows wherever it wishes; you hear the sound of it, but do not know where it came from or where it is going". This is the same spirit (wind) that the ancient biblical thinkers imagined to have hovered over the waters of chaos when time began and to have blown Earth and sky into existence.

The undefined and impersonal living force thought to be permeating all that was visible and tangible eventually became particularised. But in this process each primitive culture slowly developed its own distinctive thought world as a way of describing and explaining the environment it lived in. Most early cultures came to speak of the whole Earth as a mother and to treat it as such, for did it not supply them with the necessities of life? Using a similar trope, they referred to the sky as father—since they were unconsciously projecting human attributes onto their environment, it was only natural to project the two genders. When the sky became pictured in a yet more human form, it became the 'father in heaven' still addressed in the familiar words of the best-known Christian prayer.

One of the best examples of this myth-creating projection is found in the world view of the New Zealand Maori before they encountered Europeans. In the cycle of stories that described the Maori world, Papa (Earth mother) and Rangi (Sky father) were said to have emerged out of the womb of the primeval night. It was they who, in a very close embrace, procreated the atua (gods). One of these was Tane (the deity of the forests and birds), who forced his divine parents to separate by pushing them apart with the tall trunks of his trees and thus brought light into the world, the light that now exists between Earth and sky. According to Maori myth, this primeval event is still reflected in the falling rain and the rising morning mists, for they represent, respectively, the weeping of Rangi and Papa over their enforced separation from each other. Because of his success, Tane became acknowledged as the chief of the other gods who controlled the remaining natural phenomena between Earth and sky. One of the more important of these was Tangaroa, the god of the fish and the sea. And in addition to the gods, the Maori world view included fabulous monsters called taniwha, who lived in springs, rivers and caves.

Whereas the Maori stories reflect the bush-clad country of virginal New Zealand into which their Polynesian ancestors arrived long ago, the equivalent scenario from ancient Babylonia reflects the creative fertility of the two great rivers, Euphrates and Tigris, which ran through an otherwise-arid plain before reaching the salt water of the Persian Gulf. The famous Babylonian epic of origins, called the *Enuma Elish*, was recovered in almost complete form from the ruined Library of Ashurbanipal in Assyrian Nineveh and first published in 1876.

This epic is a complex story that combines knowledge of natural forces, items of human self-knowledge and memories of the past in an imaginative amalgam that helps to explain how the ancient Babylonians viewed the world they lived in. It began with the union of the two primeval gods, Apsu (sweet water) and Tiamat (salt water). Together they brought forth the other gods, who (as adolescents commonly do) made so much noise that Apsu wished to kill them. To prevent this, Tiamat issued a warning to Ea, the most powerful of their offspring, who put Apsu into a coma and then killed him. Then, as the chief god, Ea took Damkina to be his consort and produced a son, Marduk, who became even greater than his powerful father, for Marduk was the master of the wind, which he used to create dust storms and tornadoes. This wind so disrupted Tiamat's great body and the gods still residing within it that Tiamat responded by creating great monsters and elevating the god Kingu to the status of her new husband. Then the gods became divided; some urged Tiamat to take revenge for the death of her husband by killing Marduk, while others gave their allegiance to Marduk.

Having challenged Tiamat to combat, Marduk killed her and ripped her corpse into two parts, from which he fashioned the sky and Earth. Marduk then organised the heavenly bodies and regulated the sun and moon to form a calendar. Marduk finally killed Tiamat's second husband Kingu, and from the latter's blood created the humans to be servants of the gods. Thus, in a story much more complex than this summary suggests, we find yet another attempt to explain the origin of both the gods and humankind.

In the mythology of ancient Greece, the Earth mother was called Gaia. She emerged from primeval Chaos, gave birth to Uranus (the sky) and then mated with Uranus to produce the race of beings known as the Titans. These were led by Kronos in a revolt against Uranus, and ruled the universe until they were in turn overthrown by their offspring—Zeus and the other Olympian gods. As in the *Enuma Elish*, this complex cycle of myths that we find reflected in the epic poems of Homer and Hesiod demonstrates the rich and even lurid creativity of the human imagination.

It is worth noting that even the ancient mythologies acknowledged that 'the gods' had a beginning, for they were generated by even more primal powers such as Kronos (time), Gaia (earth), or Tiamat (salt water). But as we look back from today's vantage point and with today's knowledge of the creative human mind, we can see that the gods were created by the collective human imagination.

To be more specific, the ancients projected their own personal consciousness onto the phenomena of nature as a way of explaining the world they lived in. Although we call them myths, the stories of the gods amounted to a embryonic form of 'science' (knowledge) by which they 'knew' their world and were able to respond to it in a meaningful way; not only was human personality being projected onto natural phenomena, but so too were human sexuality and human family relationships, for these were believed to be part of the basic structure of the cosmos. In the thought-world constructed by each primitive or ancient culture, the inner experiences of humans were unconsciously being assigned to external reality. By means of stories this thought-world was shared with the other members of the community and thus passed on to the next generation.

When we hear mention of gods today we too-readily jump to the conclusion that they represent the beginnings of religion. But it is quite anachronistic and therefore misleading to use such modern terms as *religion* and *science* when discussing a cultural age where they do not belong and where they were in fact never used. No one expects to find empirical science referred to in the ancient world, and neither should they expect to hear religion spoken of. Neither is discussed in the Bible, even though that is commonly regarded as the religious book *par excellence* by Christians. In fact, on the odd occasion where the word *religion* does appear in the English Bible, it could just as well have been translated as 'superstition', for it refers to the sacrilegious practices of other nations.

For people of the ancient world, the body of cultural knowledge that was transmitted from generation to generation served as both science and religion, for it both explained the world they lived in and taught them how best to respond to it. For them it was the ultimate truth, and it was what they lived and died for.

This may be exemplified by the importance that the Maori people still attach to their cultural knowledge. Prior to the arrival of Europeans, the language of the inhabitants of New Zealand had no word corresponding to our term 'religion', but central to Maori culture was a body of traditional knowledge (now sometimes called Maoritanga) that comprised the sum total of what it means to be Maori. Of supreme importance to each individual was the duty to absorb this knowledge from the elders and then to nurture and preserve one's identity and being as

Maori. Included in this lore was knowledge of the natural world and how to use it for food, medicine, toolmaking and crafts. It also called for knowledge of and respect for the gods (atua), by which they defined and interpreted the natural world in which they lived. Above all it instilled knowledge of social values and practices as well as knowledge of their ancestors and where they came from. This knowledge was often transmitted in story form.

Some ten thousand years ago, our ancestors began to make the transition from hunter-gatherers to agriculturalists. The former were nomadic, free to move wherever they found sufficient food to sustain themselves, and this is why they eventually spread round the globe. We see similar lifestyles today in the Arab Bedouin and the Aboriginal Australians—the latter often restless to 'go walkabout'. Increasing dependence on agriculture, however, entailed a more settled life, with people living permanently in villages, and later in cities, usually near a dependable source of water.

The transition to a more settled life meant that humans began to be more actively involved in providing themselves with food. Purposeful action on the part of humans soon became as important as chance events in shaping our evolution. No longer simply harvesters of what nature provided by way of fresh meat, fruit and nuts, they were now motivated to take such a keen interest in the seasons and the vagaries of the weather that by trial and error they gradually developed agricultural skills. From sunshine and storms to the mystery of biological growth, every aspect of nature became a subject of inquiry and an area where new gods were to be discovered—that is, of course, created by human imagination. Because the gods of fertility became increasingly dominant and the seasons of the year were vitally important to agriculturists, it is not surprising that the transition from a lifeless winter to a 'springing' forth with new growth came to be explained in terms of a dying-and-rising god. The process of conceiving and naming the gods was carried out by means of language and the sharing of ideas among inquiring and creative minds.

Peering into the ancient past from a cultural context that long ago abandoned primitive polytheism, we too often fail to appreciate that, inasmuch as the gods were created by human imagination to explain natural phenomena, they were just as much concepts of primitive science as they were of primitive religion. Just as modern physicists have created such terms as electrons, quarks and black holes in order to explain natural phenomena, the ancients created terms like spirits, jinn, angels, devils and gods—all of which became important elements of the World 3 that they constructed to explain what they observed in World 1.

In the experience of the ancients, these two worlds were one and the same, and that is why the gods were very real and not the imaginary figures that we take them to be. In the ancient mind the gods were the chief points of interest in the world because they were centres of power and explanations for natural phenomena. It was through the gods that the ancients understood who they themselves were and what life was all about. This they expressed simply in stories about the gods rather than in doctrines or theological treatises. As we have noted, the stories that have survived from both pre-literate and ancient literate cultures were often very complex; they usually consisted of cycles of narratives about the same leading characters and existed in different versions. There was no original from which the later ones diverged, for they all evolved over the course of centuries.

Take for example one tiny fragment of a primitive polytheistic story that has survived in the Bible (my paraphrase):

When mankind began to multiply on the surface of the earth, and daughters were born to them, the sons of the gods saw how attractive they were and they took as their wives whomsoever they chose. That is how there came to be giants on the Earth in those days. It was because the daughters of men bore children to the sons of the gods. These were the heroes of olden times, the men of great reputation.

Today some may think it strange to find such a fragment appearing almost immediately after the story of Adam and Eve, but the biblical compilers needed it to explain why God had to send the Great Flood and wipe out the whole human race he had created and then start again with the family of Noah. These two stories reflect successive and rather different stages in the process of polytheogenesis.

The story of Adam and Eve is the later of the two and brings us close to the end of the Age of the Gods and near the advent of the monotheistic age. It contains no residual signs of the primitive gods and reflects only one deity. Yahweh (the national god of the Jews) has become identified with the more general term *elohim* (the gods of the peoples) to become the creator of everything. (This is why Bible readers have long been familiar with the translation of this combined term as "the Lord God".) It can be summarised like this:

In a desert-like world that lacked all life the Lord God fashioned the first man from the dust of the earth and transformed him into a living being by breathing the breath of life into his nostrils. After placing him in the Garden of Eden, now enriched by the plant and animal life he also made, the Lord God observed the man's loneliness and so proceeded to create a female partner for him by anaesthetizing him, removing a rib and fashioning it into a woman. The Lord God made rules for the idyllic garden life to be lived by the happy couple: they were not to eat the fruit of the tree in the middle of the garden for if they did so they would die. A serpent appeared and told them they would not die but would become like God, knowing good and evil. So they ate the fruit and they became aware of their nakedness. They tried to cover themselves and hide from God in their shame. While taking his usual constitutional walk in the cool of the day God challenged them. After hearing their excuses God passed judgment on the serpent, the woman and the man, in turn, and drove them out of the Garden Eden to work hard and suffer until they died.

No doubt this story seems very quaint to us, but, first, it reflects a linguistic logic that at one time seemed to make eminently good sense. As even a basic knowledge of the Hebrew language reveals, it is highly symbolic, for the original storyteller made a play on words: Mankind (*adam*) is made out of the ground (*adama*). Similarly the surgical operation performed by God was suggested by the way the Hebrew word for woman (*ishsha*) was clearly related to the word for man (*ish*). The woman was named Eve (a Hebrew word meaning "life-giving") because she was to be the mother of all future humankind.

Further, this story shows considerable insight into the nature of the human condition by drawing attention to our tendency to shift blame onto others, and it answers certain questions that puzzled the ancient mind: Why does such a natural process as childbirth involve so much pain? Why do snakes have no legs? Why do weeds grow in the very place where one has sown good seed? Why do people feel shame at being seen naked? Why do they have to work hard for a living instead of being able to sit back and simply enjoy the fruits of the land? Each of these represent a divine judgment for disobedience. The ancient process of storytelling combined the functions of explanation and entertainment.

Many of the ancient explanations about our origins cast a god or gods in leading roles. However else their narrators and listeners may have understood them, they were the key to knowledge and understanding. They were symbols of the mystery of life and of the transcendence of the natural world over human beings. The stories of the gods not only explained the vagaries of the natural world and thus provided a kind of primitive science, but they also explained why humans felt so dependent on their environment for both sustenance and security.

The time was to come in the evolution of human culture when the stories about the gods came to be questioned and displaced. That time is now commonly called the *Axial Period*, a term that was coined by the philosopher Karl Jaspers to refer to a relatively brief time span around 500 BCE during which radical cultural change took place. More or less simultaneously in five or six different places on the Asian continent, human culture seemed to turn as if on an axis and reject the gods on which its attention had been focused for some twenty thousand years.

Whereas Western culture numbers its years from the supposed date of Jesus' birth, the Jews number theirs—perhaps more logically—from the supposed creation of the world, and Muslims begin with the establishment of the first Islamic society. But a more natural point of demarcation in the evolution of human culture is the Axial Period—what Karen Armstrong, in her highly informative study of this cultural event, has called *The Great Transformation*. It was an era when several major ethnic communities on the Asian continent experienced increased levels of critical reflection and individual self-consciousness as a result of the inspirational impact of prophets, thinkers and seers.

It may be useful to regard the Axial Period as that stage in the process of noogenesis when the human species reached its adolescence on a grand scale. When we become adolescents we begin to think for ourselves, establish our individual identities and make independent decisions. We often challenge parental authority and begin to question various aspects of our culture that we uncritically absorbed as children. We try to sort out our beliefs and reconcile them with the flood of new knowledge and experiences that is preparing us for adulthood.

Similarly, the Axial Period was the time when the human race sought to leave behind its childhood—its sense of obligation to give blind obedience to its ethnic inheritance, including its gods—and entered its adolescent phase by seeking a new kind of human identity that superseded ethnic divisions. And thus with the Axial Period came the birth of world religions in which people transferred their ultimate allegiance from tribe, race or nation to some reality or truth that transcended ethnicity.

If we pause to evaluate the Axial Period within the much more extensive context of the evolving noosphere, we may see it as a natural culmination of what preceded it. Ever since they began the change from being hunters and gatherers to being settled agriculturalists, humans had become steadily more congregated in the great river valleys of the Nile, the Euphrates, the Indus, the Ganges and the Yangtze. Here, beginning some six thousand years ago, the earliest cities were established, and this meant that people were no longer wholly employed in growing their own food. Not only did they have time to develop new trades and skills, but they had the leisure to reflect and philosophise. With the greater diversification of occupations and the spread of knowledge, people became more sophisticated. It is no accident that the word *civilisation* derives from a term meaning 'city-dweller'.

In short, during the Axial Period polytheism was gradually superseded—but what replaced it varied considerably from place to place. In India, even though among the uneducated classes the gods lived on and indeed multiplied in number, the more perceptive thinkers marginalised them by favouring such higher and impersonal ideas as sacredness or truth. In a similar fashion the Buddha regarded the gods as irrelevant to the religious quest and replaced them with the Dharma (the practical way to find fulfilment); Confucius simply replaced them with a vague reference to "heaven".

In the Middle East, however, and largely through the efforts of the Israelite prophets, the gods were rejected and their
existence emphatically denied. Allegiance henceforth was to be
given to the 'One and Only God'. The age of polytheism (the
belief in more than one god) thus gave way to that of monotheism (the belief that there is only one god). Since the modern
global and secular age emerged out of the monotheistic culture
of the West, our story will now confine itself to the emergence
of monotheism. To this process, which may be called *monotheo-
genesis*, we now turn.

# CHAPTER 8

# Monotheogenesis

## The Emergence of
## One God

### *From 2,500 Years Ago*

Western culture has for so long been based on monotheism—the notion that the whole universe (including ourselves) was created by one supernatural, yet personal, being called God—that despite the erosive acids of modernity, our thinking remains permeated by a now-outmoded theism to a degree that we often fail to recognise. We too easily forget that as soon as we ask a question such as "Does God exist?" we betray the fact that our minds have been shaped by one of the three monotheistic traditions—Jewish, Christian or Muslim. In other cultures the question could sound puzzling. In traditional Indian culture, for example, this question might well bring forth a response like, "Which god do you mean?" In classical Chinese culture the question could not be asked, let alone answered, for the classical Chinese language had no word for *God*. In short, discussion about God usually starts off from hidden cultural premises.

Yet monotheism had a beginning, and one that is not very far distant in time. We in the West still live fairly close to the 'Age of the Gods' that preceded the monotheistic age, for much of northern Europe remained polytheistic until a thousand years ago, a fact that has left a very clear cultural residue to this day. The names of the days of the week were never Christianised and still carry the names of the gods to which they were dedicated: Sunday to the Sun-god, Monday to the Moon-god, Tuesday and

Wednesday to the Norse gods Tyr and Woden, Thursday and Friday to their fellow deities Thor and Frigg, and Saturday to the Roman god Saturn.

The question, then, is: When, where, and how did monotheism come into being? The process began in the 'civilised' city life of the Axial Period; gradually monotheism came to supersede and replace polytheism. Like most new developments in the long process of cosmic evolution, the transition took place in stages over several centuries. It is possible to sketch these stages with some confidence, for the emergence of monotheism is remarkably well documented. Indeed, it is now possible to write *A History of God*, as Karen Armstrong did in 1993. Chief among our documentary sources is the Hebrew Bible (or Old Testament), a collection of books that contain material written over a period of nearly a thousand years. Though Jews and Christians have long treated these texts as Holy Scriptures that originated with God, the study of them as documents composed by human authors reveals them to be an invaluable source for understanding how monotheism evolved out of polytheism.

They show, for example, that between polytheism and monotheism came an important stage now known as *henotheism*. This is when one god is worshipped exclusively while others are still acknowledged to exist. In the henotheistic stage each city or nation owed particular reverence and allegiance to its own patron deity or national god. Just as Athena, the Greek goddess of wisdom, was the patron goddess of Athens, so Yahweh (originally, it seems, a storm god) became the national deity of the Israelites by virtue of their belief that he rescued them from slavery in Egypt and led them to take possession of the country called Canaan, which they consequently referred to as the Land of Promise, or the Holy Land.

Throughout the classical period of ancient Israel, 1000–587 BCE, the Israelite prophets became Yahweh's spokesmen and champions (the name of the prophet Elijah, for instance, means

'Yahweh is my god'). They pleaded with the people of Israel to be faithful to their national deity Yahweh and not "to go after other gods to their own hurt". Thus they did not deny the existence of the gods of the other nations, but in effect proclaimed henotheism, enshrining it in the First Commandment: "I am Yahweh, your God, who brought you out of the land of Egypt, out of the place of enslavement. You shall have no other gods but me".

It was henotheism that gave rise to the biblical use of the phrase *god of*, as in 'the God of Abraham' or 'the God of Israel', as well as such terms as *our* God, *your* God, or *my* God. These locutions led to a subtle but very important change in the use of the word *god*. In polytheism all gods were known by their proper names: Greeks had Zeus, Apollo and others, while the Israelites had Yahweh. The word *god*, as we have seen, originated as the generic term for a class of supernatural beings and had not yet assumed the status of a proper name. This is even clearer in the Hebrew language where the word for God, *elohim*, retained its plural ending even after the advent of monotheism. In the Hebrew Bible this word refers both to the gods of the nations and to the one God of Israel, the context alone indicating which meaning is intended.

As long as henotheism persisted, the Israelite prophets fought a continual battle with the priests and prophets of the traditional gods in their attempt to stamp out polytheism. Without knowing it, they were preparing for the emergence of monotheism. Yet even more noteworthy is the fact that the advent of monotheism, destined to have such vital significance for the future of humankind, actually occurred somewhat unexpectedly and in unusual circumstances. It took place during a period of great cultural crisis, when the Jewish people were in grave danger of losing their national identity forever.

They had found themselves at the mercy of Babylonia after it had become the most powerful empire in the Middle East.

The Babylonians had not only captured the Jewish holy city of Jerusalem and destroyed their one and only temple, but had carried off the royal family, priests and aristocracy to captivity in Babylon, their own capital city on the river Euphrates. There all the leaders of Jewish society were forced to live in what promised to be a permanent exile, for Babylonian policy called for the removal of a conquered nation's natural leaders in order to prevent revolts.

The Babylonian Exile of the Jews, which they naturally lamented as a cultural catastrophe, quite unexpectedly became the occasion of their greatest period of cultural creativity. It may be compared in some respects with the way the Nazi Holocaust of the Jewish people led directly to the creation of the modern state of Israel. Out of the disaster of the Babylonian Exile there was created a set of Holy Writings (the beginnings of the Bible and a prototype of the Qur'an), the institution of the synagogue (the prototype of church and mosque) and, above all, monotheism!

To appreciate this fully, we must understand that Babylon was then one of the most culturally advanced cities in the world. As the heirs of two thousand years of Sumerian culture, the Babylonians had become skilled mathematicians, astronomers and builders. The hanging gardens of Babylon were universally hailed as one of the seven wonders of the world.

The Jewish exiles had barely adjusted to this very different lifestyle before Babylon was in turn conquered by Cyrus the Great of Persia. Not only did Cyrus create an empire larger than the Babylonian—one that lasted for two hundred years—but he introduced a style of rule that allowed some of the Jews to return to Jerusalem, rebuild their temple, and live in freedom once again. For this they hailed him as their deliverer, the Messiah (the first biblical use of this term).

Under the enlightened rule of Cyrus, the innovative thinking of the Persian prophet Zarathustra became part of the rapidly changing culture of Babylon. This man had already been successful in replacing Persian polytheism with a morally based monotheism focused on Ahura Mazda, 'Lord of Light'. The Jewish exiles found themselves in a cultural maelstrom; to ensure their ethnic survival they needed to sort out and reinterpret their own traditions in the light of all the new knowledge and challenging experiences they were encountering. We know little of how they went about doing this, but we do know that they went into exile as henotheists and returned to their Holy Land as monotheists.

Possibly the earliest extant declaration of pure and undiluted monotheism is to be found in some words (now included in the book of Isaiah) from an anonymous Jewish prophet of the sixth century BCE. He announced to the Jewish people an oracle from Yahweh their god: "Turn to me and be saved . . . for I am God, and there is no other." This marks the point at which Yahweh, previously regarded as the national god of the Jewish people, is dramatically elevated to the status of the one and only real deity.

In the opening chapter of Genesis we find the earliest expression of the notion that God is the creative source of all that exists, the maker of Heaven and Earth, and hence the ultimate explanation of everything and the key to the meaning of human existence. Instead of dismissing this document as hopelessly outmoded in the light of modern science (as we have been doing for the last two hundred years), we should evaluate it in its original cultural context. When we do, it turns out to be one of the most succinct and imaginative documents of origins ever composed. We may even marvel at its comprehensive simplicity, for nothing else in the world of that time can match it.

We should further note that the Jewish composers of this document do not even mention the name of their national god, Yahweh, the appellation used in the First Commandment, a statement that exemplified henotheism. In this document the creator of the universe is not referred to as Yahweh, or even as "the God of Israel", but simply as "God". It is here that the word *God* assumed the status of a proper name and perhaps for the first time!

Pure monotheism had at last emerged in affirming that God alone is the creator of everything. But what was already there for God to work with? Only a watery shapeless void—perhaps the writer's imaginative attempt to evoke nothingness. (Have you ever tried to imagine "nothing"? It is more difficult than you may think.) But it may also preserve a remnant from the Babylonian *Enuma Elish*, referred to in the previous chapter.

Thereafter the Genesis document relates in a very orderly and simple fashion the creation of all known entities in an order of increasing complexity. Each component of the universe is allotted to one of the first six days of time. We may find it illogical to have the sun and moon created on the fourth day, since they mark the passing of time, but this apparent illogicality may have been intentional. The sun and the moon, formerly worshipped as gods in the days of polytheism, were effectively devalued and 'put in their place' by being relegated to the fourth day.

So the universe was created at the personal command of God and in the following order:

Day 1: Light

Day 2: Space (separating Earth and sky)

Day 3: Dry land and vegetation of all kinds

Day 4: The heavenly bodies (sun, moon and stars)

Day 5: Birds and fish

Day 6: Animals (including humankind)

Day 7: The sabbath (day of rest)

This climax shows how the priestly composers were intent on citing divine authority for the strict observance of every seventh day as a day of cessation from all work. This is because the Jewish observance of the sabbath, along with the practice of male circumcision, had become the primary means of distinguishing the Jews from all other inhabitants in Babylon, and thus assisted their ethnic survival. It may seem ironic that this inspiring account of creation was in part a by-product of their desire for self-preservation, but are not such unintended consequences quite typical of the whole story of cosmogenesis? As it turns out, this document came to exert a cultural influence far beyond that of merely validating the sabbath.

First, we should note that whereas previous cultural accounts of origins had taken the form of a story, this statement displays the style of a thesis—a detailed set of assertions. In its own time it was a 'Theory of Everything', to borrow a phrase used by modern scientists. In fact, there had never before been a theory known to equal it for simplicity and comprehensiveness, and that remained so for two millennia. In the days before empirical science a thesis did not need to be proved by supporting evidence; it simply had to make sense and be convincing. This thesis was eminently convincing and remained so right up until modern times. It even passed the test of Occam's razor, which asserts that the simplest credible theory is to be preferred.

Further, this theory of creation provided a clear description of what was meant by the word *God* and is one reason why monotheism has lasted up until the present. What was so essentially new about monotheism was not the *theism*, for it shared that with polytheism. What was new was the *mono*; the notion of God as the sole Creator provided unity and meaning to everything, unifying everything that transcends us and making all other gods superfluous. It follows from there being only one Creator that we live in a *uni*verse and not a *multi*verse. At its most basic level the word *God* symbolises the unity of the universe. It is little wonder that monotheism became the foundation of two great civilisations, Christian and Islamic. What is more, as we shall see in the next chapter, it provided the basis for the rise of empirical science.

But first we must look at the wider world into which monotheism was about to spread from its originating point in ancient
Babylon. This was not the only place where the transition from
polytheism to monotheism showed signs of taking place. A little
earlier in Egypt the Pharaoh Akhenaten (died c. 1336 BCE) promoted the worship of the sun-god Aten and used his royal power
and authority to banish the traditional gods of Egyptian polytheism. Some have hailed this as the first appearance of monotheism, but it had no lasting significance, for upon Akhenaten's
death the priests not only rejected it but erased the Pharaoh's
efforts to promote it.

It was a different story in Greece, where the philosophers tried
to see beyond the traditional Greek gods, so vividly portrayed
by Homer, in order to reach the essence of deity. But the *theos*
they spoke of, though a unity, was an impersonal entity conceived somewhat differently by each philosopher. Plato's *theos*,
for example, was the eternal 'form', or timeless idea, behind the
humanly conceived *theoi*, the gods of Olympus. Whereas the
traditional Greek gods were anthropomorphic and dispensable,
Plato's *theos* was impersonal and eternal.

Aristotle rejected Plato's timeless 'forms' because he was more
interested in understanding the physical world. Being a logician,
he traced everything in the cosmos back to one initial cause, the
Prime Mover, which he understood to be eternal, immaterial and
unchangeable. That was his *theos*. The Stoics, in turn, conceived
*theos* to be the principle of rationality and order which pervaded
all things. For them *theos* was *logos* (reason), the soul of the natural world—an idea that we unexpectedly find indelibly expressed
in the opening words of St John's Gospel: "In the beginning
was the Logos, and the Logos was with God, and the Logos was
God. All things were made by him."

As this example illustrates, the thinking of the Greek philosophers began early on to penetrate the Christian understanding
of God as Christianity moved out of its Jewish origins and into

the Graeco-Roman world. This influence was even more pronounced in the Neoplatonist philosopher and mystic, Plotinus (c. 205–270 CE), who in turn had a great influence on St Augustine of Hippo. Thus, the notion of God that became so basic to the world view of Christian culture owes almost as much to the Greek philosophers as it does to the creative Jewish thinkers in ancient Babylon.

Indeed, there has never been a single exclusive and definitive way of understanding God. Although often asserted to be by his very nature unchangeable ("the same yesterday, today and forever"), God has been imagined and understood in ever-changing ways, which is why God has a history that can be told.

Monotheism spread because it was a much simpler explanation of the observable phenomena of world than that offered by any form of polytheism. Yet from the beginning pure monotheism encountered problems, the first having to do with God's morality. The first prophetic voice to enunciate monotheism portrayed God as one who declared, "I form light and I create darkness. I bring health and *I create evil*. I do all these things" (emphasis mine). If God created everything, and continued to be responsible for everything, he was clearly the author of evil. It was only a century or so after the emergence of monotheism that the author of the book of Job challenged the moral justice of the one Creator God. Over the intervening centuries many attempts have been made to solve this problem, among them being the interpretation of evil as a form of divine punishment. The fact that it is still common to employ this explanation shows how persistent this moral dilemma posed by monotheism is.

Another attempted solution was the creation and subsequent evolution of the supernatural figure of Satan. This let God off the hook by transferring the blame for all evil to Satan, but by the Middle Ages Satan had become such a powerful figure in popular imagination that the resulting dualism in the supernatural world threatened the central and unifying essence of monotheism. The as-yet unsolved moral problem of undeserved suffering has proved to be the Achilles' heel of monotheism. As late as the eighteenth century the philosopher Leibniz coined the word *theodicy* for this perennial conundrum.

Monotheism experienced its most radical changes in the rise and evolution of Christianity and its metamorphosis into the Holy Trinity. When monotheism first emerged, the Jews had little trouble uniting their traditions about their national god Yahweh with the new declaration of the one and only Creator God; indeed, it allowed them to see themselves as 'the chosen people'. This synthesis was carried through in Babylon in a remarkably short time and resulted in the compilation of the Torah (Five Books of Moses), where the seams of the combining process are still clearly visible to scholars. Christians, however, experienced extreme difficulty in attempting to blend Jewish monotheism with their newly proclaimed allegiance to the man Jesus of Nazareth, and it took them some four centuries to achieve a solution.

First came the concept of the incarnation, a notion so old that it appears at the end of the first century in St John's Gospel. To explain the mantle of divinity with which pious Christian imagination had begun to clothe their Lord, it was claimed that Jesus was the incarnation of God, "God in human flesh", but this raised the problem of how Jesus could be both wholly human and wholly divine at the same time. If he was divine, how and why did he pray to God? The problem gave rise to a variety of attempted solutions, all of which were subsequently declared to be heretical.

The lengthy and often fierce debates were concluded (though by no means unanimously) at the ecumenical Council of Chalcedon in 451 CE. In the course of four centuries Christian thinkers came to understand God as a Holy Trinity consisting of Father, Son and Holy Spirit. They constructed this model in order to preserve monotheism while at the same time affirming both the humanity and divinity of Jesus and establishing the authority of the divine spirit within their community. The Holy Trinity was defined in philosophical terms as three *hypostases* in one *ousia* (the Greek version) or three *personae* in one *substantia* (the Latin version).

This complex verbal symbol was a far cry from both the god of Plato and the god of the Israelite prophets. It is little wonder that a substantial minority of Christians rejected the final formula even in the fifth century. Nor is it any surprise that when Muhammad, the Arab prophet and founder of Islam, encountered Christianity he concluded that Christians were no longer monotheists like the Jews but had become tritheists, people who acknowledged three gods. But even then, he got them wrong: he assumed (perhaps after observing the icons that appeared in the Byzantine churches) that the Trinity consisted of the Father, Son and Virgin Mary. In any case, since both Jews and Christians had come to believe that monotheism originated with the Jewish patriarch Abraham, Muhammad felt called to restore the pure monotheism of Abraham.

It is a moot point whether even Christian theologians really understood the verbal explanation of the Holy Trinity; it is certain that ordinary people did not, even though they kept reciting it in their creeds and liturgies. This helps to explain why Islam moved so quickly as it spread out of Arabia into the Byzantine world: In their relationships with Christians, Muslims saw their mission as one of restoring the pristine monotheism (by then thought to go back to Abraham) that was preserved by the Jews but abandoned or confusingly distorted by the Christians.

Just as species diversify and languages multiply in biological evolution, so do religions diversify in cultural evolution. Monotheism diverged into the three Abrahamic religious traditions that exist to this day: Jews, Christians and Muslims all claim to be monotheists and assert that there is only one God. But are they referring to the same God? For Jews, God is the One who delivered their forbears from slavery, gave them the land of Israel and providentially continues to preserve their identity. For Christians, God is the One who became incarnate in Jesus Christ, sacrificed himself for sins of humankind and rose again to be the living Head of the Church. For Muslims, God is the One who appointed Muhammad as the prophet to restore monotheism and delivered the Qur'an to him through the angel Gabriel. Since 'God' can only be explained in descriptive words or further teaching and, as I have just shown, their teachings about God differs significantly, it is clear that Jews, Christians and Muslims do not acknowledge the same God. What they have in common is their conviction that there is only one God, the Creator of the universe and of us.

It is this understanding of the term *God*, first clearly enunciated in Genesis, which is usually presupposed in any contemporary discussion about God. When monotheism first emerged, its chief competition was some form of polytheism, and the ease with which this new belief system met the simplicity criterion required by Occam's razor made it preferable. Over time it came to be accepted as self-evidently true and did not require proof or verification. This still remains the case among traditional Jews, Christians and Muslims.

Therefore, when theologians like Anselm (1033–1109) and Aquinas (1225–74) began to present proofs for the existence of God, their arguments were essentially superfluous. Today all such 'proofs' are regarded as invalid, but their popularity over many centuries has offered an ironic indication of how human reason would eventually erode confidence in the reality of God.

The growing appeal of rational thinking was stimulated in part by the impact of Islamic thought and culture on the West in the twelfth century. Over the course of some three centuries, Islamic civilisation became the most advanced in the world and had spread as far as the Iberian peninsula. Since the Islamic military advance into France had been halted, Spain became the place from which some elements of Muslim thought and culture began to spread northward. In particular, it was through the Muslims that Christendom came to know of the Greek philosopher Aristotle, a seminal thinker long lost to the West.

The Muslims have a dictum that runs like this:

God is on his throne.
The fact is known.
The manner of it is unknown.
Belief in it is imperative.
Inquiry about it is heresy.

In spite of the dire consequences, however, some Islamic thinkers did dare to inquire. One of them was Averroës (1126–98), the most famous of the Islamic philosophers. He lived in Spain, where he became the authoritative interpreter of the books of Aristotle. Averroës was a rationalist philosopher who soon came to be known and studied in Christian Europe. Indeed, the arrival of Aristotle's philosophy of nature into the newly established Christian universities of Europe caused quite a stir, for it often conflicted with those elements of orthodox Christian thinking that had been deeply influenced by Plato.

The work of Averroës introduced into Western academia both a greater dependence on rational thought (as opposed to what was considered to be divinely revealed) and a greater attention to the physical and natural world. It fell first to Albertus (1200–1280) and then to his student Thomas Aquinas (c. 1225–74) to restore some consistency and unity to Christian teaching. They did this by embracing Aristotle's philosophy of nature and blending it with traditional Christian teachings.

By the High Middle Ages the prevailing world view of the West had become strongly dualistic: people imagined an invisible but perfect heavenly world above, and a disordered and sin-stricken world here below. Both the writings and the art of the period make it clear that the attention of the Western mind was chiefly focused on the upper world, for which life on Earth was believed to be both the preparation and a testing time. Visual artists used their craft to portray biblical scenes and heavenly places, finding nothing to inspire them in this natural world of bleak deserts and craggy mountains. After all, had they not been taught to see ours as a fallen world, destined to soon pass away? With the gift of hindsight we can now surmise that it was the exaggerated reality and importance given to this upper world that led to its eventual collapse.

The Thomistic synthesis involved the creation of a new term, *supernaturalis*, one not found in the classical Latin of ancient Rome. By describing reality as consisting of two different spheres, the natural and the supernatural, this new term acknowledged both Aristotle's focus on nature (to be studied by human methods) and the existence of a spiritual world (to be studied by appeal to divine revelation). Aquinas' blend of two traditions provided the framework for Roman Catholic doctrine right up until the Second Vatican Council (1962–65), but it was by no means universally accepted. It now appears that the two-world solution by which Aquinas restored some harmony to Christian intellectualism contained the seeds of its own destruction, and when it collapsed, it brought about the demise of God as well! (I have written at some length about this in *Christianity without God* [2002].)

Even before Aquinas, Francis of Assisi (1181–1226) had begun to view the world differently. Indeed, his love of nature was to make him one of the most honoured saints of these modern ecological times. Francis placed great emphasis on living a simple life in tune with nature, and he founded the Franciscan Order of Friars to put this view into practice. The Franciscans were somewhat critical of the Thomistic synthesis from the start, although it was strongly defended and promoted by the Roman Catholic Order of Dominicans.

One of the Franciscans was William of Occam (c. 1300–1349), a vigorous and independent thinker in whom we find the rationalist influence of Averroës taking root. He was largely responsible for the spread of a new philosophy that became known as *nominalism*. The prevailing way of thinking had originated with Plato, who had maintained that eternal reality is to be found not in the physical objects we see but in the forms or universal ideas they reflect. In direct opposition to this, the nominalists contended that only particular things really exist, and these simply exemplify the universals. Universals, they said, are simply concepts or names (*nomina*) that have been invented by the human mind.

Occam himself did not draw the conclusion (as we may) that since 'God' evolved from 'the gods', a generic term invented by the human mind that which it names has no real existence, but he did assert that mankind has no knowledge of God, even by intuition. It is faith alone, not reason, that provides the basis for all theological assertions about God. In contrast with Aquinas, he believed it impossible to present rational and cogent proofs of the existence of God. By driving this wedge between philosophy and theology, Occam destroyed the Thomistic synthesis. For him, theology and philosophy were two quite separate intellectual disciplines: Theology explores and expounds what has been divinely revealed and can be apprehended only by faith, and philosophy (out of which all the sciences were eventually to emerge) explores those aspects of reality that can be examined and understood by observation and human reason.

Although Occam did not draw the contrasts as sharply as this brief sketch has done, even fourteenth-century thinkers began to sense that they were at a crossroads and that nominalism was leading them in a new and different direction. It was already being referred to as the *via moderna* (the modern way) in contrast with the *via antiqua* (the ancient way). Thus, in his own day Occam's ideas were already recognised as a threat to Christian orthodoxy; it should therefore come as no surprise to learn that he was excommunicated. In spite of his fate, nominalism began to spread through the universities. Although this had the effect of causing philosophy to become a lay pursuit rather than a clerical one, it would be some centuries before the divorce between theology and philosophy was complete.

In the long run nominalism led to the realisation that, far from being the name of an objective, though unseen, spiritual being, *God* is in fact a humanly created word that originated in the distant past and has evolved to become an important bearer of meaning—the oneness of all reality. To be sure, theological statements consist of nothing more than words, and some people understandably find them to be meaningless gobbledegook. On the other hand, as is true of poetry, theological statements are attempts to express human experience, particularly in the quest for meaning and purpose in life. In this respect the word *God* does partake of a kind of reality, but its reality is of the order found in World 3.

The spread of nominalism was one factor that helped to promote the Protestant Reformation, for even Martin Luther acknowledged himself to be a nominalist. The Reformation fragmented the Western Church, which had already become separated from the Eastern Orthodox Church, the Egyptian Copts, and the Far Eastern Nestorians. Just as the institutional church divided into denominations (another example of the evolutionary tendency to diversify), so also the very word *God* began to show a diversity of connotations, as we can see from the variety of terms that came into use.

What had long been the traditional understanding of God (a view that owes much to the Bible) was being referred to as *theism*: Theism assumes God to be the personal name of the supernatural spiritual being who created the world and continues to oversee its affairs, intervening from time to time with miraculous results. Being personal, this God enters into personal relationships with humans, who, the Bible tells us, are made in his image. Christian orthodoxy still strongly affirms and defends theism. As one of the essential tests of orthodoxy, Evangelical Christians ask, "Do you believe in a *personal* God?"

*Atheism* is a word that originated in the ancient world to refer to those who gave no credence to the Greek and Roman gods, and it should not surprise us that the first Christians were dubbed atheists. The modern use of the term dates from the eighteenth century and refers to people who deny that any reality can be ascribed to the concept of a deity. Because the word carries many negative associations, some prefer to speak of themselves as non-theists, even though etymologically the two words mean the same thing.

As early as the seventeenth century the Jewish philosopher Spinoza (1632–77) had broadened the meaning of 'God' to include all the forces at work in the natural world. Thus he spoke of "God or Nature" and dispensed with the idea of God as a personal and caring being. This view became known as *pantheism* (God is everything) and, though strongly rejected by most people at the time, it surfaced in a variety of twentieth-century forms expounded by such thinkers as Paul Tillich and Pierre Teilhard de Chardin. Some ask why we bother to use a separate name for something that can be identified with the natural order.

The understanding of God espoused by the leaders of the Enlightenment was known as *deism*, a term that owes more to Aristotle than to the Bible. Deists rejected theism in that they not only repudiated all the personal attributes ascribed to God but also rejected the idea of miracles as divine or supernatural interventions into nature. They retained the word *God*, however, to refer to the creator of the universe. Deism appealed to critical thinkers during the rise of modern science and it is still widespread at a popular level, often among people who have never even heard the word. When modern physicists such as Albert Einstein, Fred Hoyle, Stephen Hawking and Paul Davies use the term 'God', they are usually thinking of an infinite intelligence behind or within the creativity of the universe, and this view is much closer to pantheism and deism than to theism. Advances in the natural sciences seemed to leave fewer places in which to discover the activity of God, and accordingly there arose a new label for the deity: 'God of the gaps'. In any case, as more and more people came to see divine activity restricted to the original creation, God was being eased more and more out of the world of here and now.

By the mid-eighteenth century a few brave souls dared to confess themselves to be atheists. It took courage to do so, for the word carried such stigma that for fear of social ostracism most unbelievers were reluctant to follow suit until well into the twentieth century. The scientist T H Huxley (1825–95), a great admirer of Charles Darwin, coined the term *agnostic* for himself, indicating that he did not profess to know whether God exists or not; many still prefer this term.

Paradoxical as it may seem, the nineteenth century marked both the time when Christian monotheism was beginning to erode and the time when the Christian tradition came nearest to achieving its goal of becoming a global movement. Its unprecedented expansion around the world due to missionary endeavours led Christian historians to refer to it as "the Great Century". Triumphant Christian leaders entered the twentieth century promoting the slogan "The evangelization of the world in our generation".

Yet it was near the end of that very century that Friedrich Nietzsche (1844–1900), in a striking parable about a madman, declared God to be dead. He recognised that the basic concept on which Western civilisation had been founded was slowly being eroded and marginalised, and he wanted to warn people of the *nihilism* (the total rejection of current religious beliefs and moral principles) that he believed this erosion would lead to. As he saw it, human existence would become empty of meaning, purpose and any lasting value. In one sense he was right, for God had long served as the symbol for what created and held the universe together, as well as supplying it with purpose and meaning; without such a basic symbol, everything we value becomes transitory, short-lived and ephemeral. Most people dismissed his message at the time and many still do, preferring to believe there is an ultimate meaning to life if only you can find it.

But Nietzsche proved to be more of a prophet than most realised at the time, for during the twentieth century the churches began to empty, fewer people trained for the clergy and Christian missionary endeavours ground to a halt. Then, sixty years after Nietzsche, a few leading theologians took up his cry and declared God to be dead. They were speaking metaphorically, for God had never been alive except in the collective human imagination (the contents of which, as we have seen, belong to World 3, the world of human thought).

Further, it is now abundantly clear that the idea of God no longer functions as it used to in the collective mind of Western society. John Macquarrie (1919–2007), a professor of theology at Oxford, put it well in his Gifford lectures (*In Search of Deity*, 1983–84):

> There was a time in Western society when "God" was an essential part of the everyday vocabulary. . . . But in the West and among educated people throughout the world, this kind of God-talk has virtually ceased. . . . People once knew, or thought they knew, what they meant when they spoke of God, and they spoke of him often. Now in the course of the day's business we may not mention him at all. The name of God seems to have been retired from everyday discourse.

Indeed, God is no longer being appealed to publicly in the way he used to be in what was once Christendom. The reality of God as Creator and divine Superintendent of Earth is no longer self-evident. Opinion polls find that the majority of people still say that they believe in God but such belief has now become a matter of personal choice or conviction. God has been privatised and is no longer a public figure. When thought of at all, God is conceived in a great variety of ways. The traditional model of God is still acknowledged in churches and other religious communities, but this very fact makes them appear to the outside observer like little islands of the past that have become frozen in time, like the Amish people in United States or the Hasidic Jews in Jerusalem. As for the public acknowledgment of God, it is confined to a few well-known phrases that have become stereotyped in language and most often heard in the exclaimation, "Oh, my God!" The Age of God, however vibrant it remains in Jewish, Christian and Muslim fundamentalism, is drawing to a close.

That is not wholly surprising, for whatever has a beginning also has an end. In other words, as we have noted, God has a history. Just as God emerged out of the demise of the gods, so the traditional God has in turn been superseded. As the age of monotheism has drawn nearer to its end, it has begun to metamorphose into the age of scientifically-based *humanism*. The point of no return in this transition was the Enlightenment, which marked a second Axial Period. As the First Axial Period can now be seen as the time when the human race entered its adolescence, so the Second Axial Period may be viewed as the time when humanity entered its adulthood.

Since the Enlightenment we humans have been learning, through the advent of the sciences, how we evolved slowly from the dust of Earth and we have been coming to appreciate the tiny place we occupy in this awe-inspiring universe. Now we are beginning to discover the grave responsibilities we must assume—not only for other forms of life but for the future of the human race itself. In the Age of God these responsibilities were believed to be matters of divine prerogative; in this age and the age to come they are becoming ours. It is to the coming of the present humanistic age we now turn.

# Homogenesis

## 'God' Comes Down to Earth in Humankind

### *From 250 Years Ago*

A prominent feature of evolution is that change occurs in small, incremental steps punctuated by sudden jumps. Thus the beginning of each age in cultural evolution has generally been quite imperceptible, just as the first signs of life were in the story of biogenesis. But the transition from the age of monotheism to the humanistic age that we have now entered is more difficult to discern simply because we live so close to it—so close that many people refuse even to acknowledge that it has been occurring. I have traced the rise and nature of this new cultural age more fully in *Faith's New Age* (1980), later slightly revised and published as *Christian Faith at the Crossroads* (2001). I would have preferred to call the book *God Comes Down to Earth*, and I have adapted that for the title of the present chapter.

The Enlightenment may be regarded as the chief threshold marking the transition from the monotheistic age to the modern humanistic (or secular) age, but the two remain so intertwined that the challenge of disentangling them raises considerable debate. Nevertheless, the humanistic age is becoming global in a way the monotheistic age never quite managed. Whereas the last chapter concentrated on the rise and decline of the monotheistic age, this chapter covers much the same historical period but focuses on the emerging humanistic age.

For its earliest signs we must go as far back as the First Axial Period, for at the very same time as monotheism was developing among the Jews in Babylon, the first hints of a yet-to-be-developed humanism were beginning to appear elsewhere. By humanism I mean that philosophy of life that exemplifies the maxim of the pre-Socratic philosopher Protagoras (c. 485–411 BCE): "man is the measure of all things". Humanism focuses attention on the human condition to the virtual exclusion of all supposed non-human forces and especially supernatural ones.

Confucius, for example, while never completely abandoning the thought of non-human phenomena, marginalised any supposed supernatural forces and referred to them vaguely as "heaven", while he concentrated his attention on the best way to nurture the ideal human being. Though he saw himself simply as a follower of ideals that had come to the fore during the centuries before him, his teachings both unified and magnified them. They struck such a responsive chord among his fellows that by 100 BCE he was acclaimed the First Teacher, the Wisest of the Wise. Though he came to be highly revered, he was never deified. His teachings, preserved as a collection of short sayings, *The Analects of Confucius*, become the basis of Chinese education, and for more than two thousand years generations of students learned them by rote. Confucius provided the ethical base for classical Chinese culture, while Buddhism and Taoism provided the metaphysical dimensions.

In a somewhat similar way and at about the same time, Gautama the Buddha pioneered a basically humanist path for his followers to walk. Though he never questioned the reality of the gods worshipped by his fellow Hindus, he regarded their existence as irrelevant to his goal, which was to find the best way of dealing with the suffering that he believed lay at the root of all human unhappiness. He taught that this tragic condition so permeates our lives that we are in danger of being overwhelmed by it. To be completely delivered from suffering it is necessary to find a way of escaping from the wheel of birth, death and rebirth, the continual succession from one life to another that was an all-but-universal belief throughout Hindu culture. It was by his own unaided efforts that Gautama reached what came to be called the state of Enlightenment. This in turn enabled him to devise a way of living—the Eightfold Path—that would lead to the cessation of suffering. This entirely individual or 'self-help' solution to the problems of life as he saw them showed people how to work out their own salvation without resort to any external aid, human or superhuman. Therefore, in its original form Buddhism must be judged to be wholly humanistic, even though in most of its later developments throughout Asia it took on many of the trappings of supernatural beliefs and practices.

Also in the First Axial Period, and almost contemporaneously with Confucius and Gautama, the first Greek philosophers were taking early steps towards a more humanistic world. Xenophanes (c. 570–475 BCE) scorned the ancient Greek gods of Olympus and asserted that they were so anthropomorphic (made in the image of humans) as to suggest that if animals could create deities, then horses would worship gods shaped like horses. He went on to argue for a single, motionless, non-anthropomorphic god who controlled everything simply by the power of thought.

The Stoics, writing in a cultural context greatly influenced by Plato's concentration on the essence (or idea) of deity, defined God as "a rational spirit having itself no shape but making itself into all things". By focusing attention on reason as the great creative source, the Stoics were virtually but unknowingly acknowledging humans as the creators of the concept of God, since reason must be acknowledged as one of the skills of the human mind.

The Roman poet and philosopher Lucretius (99–55 BCE), in his only surviving work *De Rerum Natura* (On the Nature of Things), denied that the universe was created by the gods or in accordance with any prior design, but that it existed due to the random and accidental collision of atoms. He denounced religious belief as the source of much misery and human wickedness. He may therefore be called not only an early forerunner of modern humanism but even one of its nurturers, for a long-lost copy of his book was discovered in a German monastery in 1417 and it played an important role in fostering humanism during the Enlightenment.

But the Graeco-Roman move towards humanism was largely lost sight of as powerful Christian monotheism spread through Europe, filling the cultural vacuum left by the fall of Rome and the loss of its intellectual heritage. The early seeds of humanism were soon buried beneath the metaphysical superstructure constructed by the evolving Christian world view, just as the strongly humanistic tendencies in the teachings of both Confucius and the Buddha failed to evolve further in the mystical atmosphere of the Orient. In India, Buddhism was reabsorbed into Hinduism; in China, Confucianism merged with Buddhism and Taoism to form a trio of traditions that operated in reasonable harmony in classical Chinese culture.

If such humanistic tendencies appeared as early as the First Axial Period in some places, why did they take so long to come to fruition? And why did the humanistic age eventually emerge not out of Confucian China or the Buddhist Orient, but out of the strongly monotheistic Christian West? There may be no definitive answers to these questions, for the process of evolution, whether in biology or culture, proceeds more by accident than by design and is full of surprises. Nor does human history move in a straight line, but the indubitable fact that the modern world emerged out of Western Christendom strongly suggests that Judeo-Christian culture may have provided the most suitable seedbed to nurture the seeds of the humanistic age that is now spreading round the globe.

To explore this possibility let us start with the Jewish tradition as recorded in the Old Testament. The book of Job belongs to that part of the Old Testament that is now known as the Wisdom stream of ancient Jewish thought, a tradition quite independent of the Mosaic and Prophetic streams. Whereas the latter two became increasingly dominant in Judaism, Christianity and Islam, the Wisdom stream became marginalised and obscured. With the advent of modern biblical scholarship it has gained increasing attention and is now referred to as Hebrew humanism.

The Wisdom stream came from the Jewish sages who, like Confucius and the Buddha, focused on the daily life of the human individual. Moreover, they were interested in their fellow Jews not as a chosen people but simply as human beings. The sages believed that humans should not seek help outside of themselves or expect others to provide for them; rather, they contended, humans had it within themselves to achieve prosperity and peace and to make a success of life. Like Confucius and the Buddha, the Jewish sages encouraged personal effort and individual initiative, calling people to be diligent and industrious. The sages did not focus on the sovereignty of God as the Jewish priests and prophets did (or as traditional Christians have done). For the sages, *God* was not conceived personally as 'the Lord of history' but was a symbolic way of referring to the cosmic order, and gathered up under one term all that humans must learn to accept about the way the world operates. For them it largely represented what we today call the laws of nature.

The sages exhorted people to take full responsibility for their own lives and discouraged them from expecting God to deliver them from suffering by miraculous interventions. They taught that since most things in life cannot be changed, people have to learn how to make the most of the choices that are open to them. The sages relegated God to the role of an impersonal creative force that had shaped the world and made it as it was, but showed no interest in human affairs. Since nature could not be manipulated by humans, acceptance of and reverence for its structure was the beginning of wisdom.

Because much of what the sages stood for is clearly expressed in the book of Ecclesiastes, it should come as no surprise that after many centuries of neglect this book is being quoted more and more in our humanistic age. Ecclesiastes complained that wherever he turned, he could find nothing permanent. Trying to create something lasting, he said, is as futile as chasing after the wind, for everything eventually passes away and disappears. He observed that what happens to people in life is often unfair, and concluded that human existence is utterly devoid of any purpose; things happen mostly by chance. Although he did not use the following analogy, his essential message can be expressed thus: since life is like a trip to nowhere, we must forget about the destination and simply enjoy the journey. (I have written more fully about the remarkable contribution of Ecclesiastes in *Such is Life!* (2010).)

Ecclesiastes asked a number of such basic questions as: What does it mean to be human? In the absence of any certainty or permanence, how can the pursuit of wisdom enable us to get the best out of life? But he did not see wisdom as an unchanging body of knowledge that can be passed on to others and says we must learn to walk the path of wisdom for ourselves rather than expect to receive it from others in a ready-made packet. Thus, like Confucius and the Buddha, this Jewish sage was blazing the humanist trail.

Three centuries after Ecclesiastes an even greater Jewish sage arose; we know him as Jesus of Nazareth. Many respected modern scholars (such as Robert Funk, Marcus Borg and Dominic Crossan) have discovered that once we strip away the mantle of supernatural divinity with which later tradition so quickly clothed Jesus, the outlines of the wholly human figure who emerges reveal him to be more of a sage than a prophet, priest or king. These scholars claim to have uncovered what they call the 'footprints and the voiceprints' of a 'Jesus we never knew' and who has been largely hidden for nearly two thousand years.

Their findings have led to the conclusion that the historical Jewish teacher who sparked the rise of Christianity was even more of a secularist—that is, a this-worldly figure—than had long been thought. Coinciding with the crumbling of dogmatic Christian doctrine, this has been a very timely discovery.

In the view of these scholars, the most genuine traces of what Jesus taught are to be found in his parables and in his short aphorisms, such as those collected in the well-known Sermon on the Mount. We may be surprised to discover that in the most strongly authenticated sayings of Jesus, he never referred to himself at all, never claimed to be the messiah, never spoke of his coming sufferings and death and did not predict the end of the world.

Further, in his parables and aphorisms, Jesus said little about God; he talked about the *Kingdom* of God, often beginning with the words "The Kingdom of God is like..." By 'Kingdom of God' Jesus did not mean the political restoration of the kingdom of David for which the Jewish zealots were so ready to fight. Rather, he pointed to a new kind of human community, a new way of living together in the here and now, one based on mutual love for fellow humans, irrespective of race, class, gender and age. He went so far as to say we should love our enemies.

In his parables, Jesus spoke of everyday things in the lives of his audience—a robbery on a lonely road, a wayward son, labourers in a vineyard, the baking of bread, the sowing of seed, a lost coin, the growth of a mustard seed, the discovery of hidden treasure, a lost sheep, a dinner party, rich farmers and money held in trust. These were not sacred or 'religious' topics but very worldly ones. On the basis of his parables and aphorisms it appears that Jesus rarely spoke about religion at all, at least not as that term is commonly understood today.

This has led Don Cupitt, one of today's most perceptive theologians, to conclude in his recent *Jesus and Philosophy* that "Jesus was an almost secular teacher, whose teaching was entirely concerned with attempting by all means to persuade his hearers to drop everything and commit themselves wholeheartedly to a quite new moral world." This means that if Jesus is to be regarded as the founder of a new religion, then instead of being named Christianity, that religion might be better called the 'New Religion of Life'.

Thus, the sage Jesus complemented the sage Ecclesiastes in a most important and positive way, for whereas Ecclesiastes focused on the world's lack of any clear purpose and on the fact that nothing lasts, Jesus focused on what we *can* make of life in the here and now while it does last. Indeed, it might reasonably be argued that Jesus created a new religion based on love for one another. He encouraged people to focus their attention on the simple but very difficult practice of serving one another in such a way as to create a new kind of loving community, the one he called the Kingdom of God.

Jesus cannot be said to have been what we today would call a humanist, for as a first-century Jew he undoubtedly accepted the overriding reality of God. Yet he clearly placed concern for human welfare ahead of meticulous observance of the supposedly divine law—we need only recall his dictum, "The Sabbath was made for man, not man for the Sabbath". Beyond doubt, then, the seeds of humanism can readily be discerned in the teachings of the Jewish sages, especially in Ecclesiastes and Jesus.

But if the seeds of humanism have long existed in both Judaism and Christianity, why were they so long marginalised, ignored and even hidden from view? The reason Jesus' humanist teachings failed to produce the new kind of society he envisaged is that his followers allowed themselves to be diverted by his unfortunate death. To be sure, he attracted a popular following who found his message fresh, attractive and down-to-earth, but his repeated invocation of a new kind of kingdom led the Roman Governor Pilate, ever fearful of a popular uprising against Roman authority, to regard Jesus as a threat to the *Pax Romana*, and to order his execution by the prescribed method: crucifixion.

The sudden and brutal execution of Jesus prompted his shattered followers to seek an explanation for what they saw as a meaningless tragedy. Within a very short time, and partly because of the manner of his death, some of his followers saw Jesus as a martyr—not an uncommon phenomenon in religious history. Within decades they had deified him; the humanist teaching of Jesus, so highly prized and committed to memory by his very first followers, soon became overshadowed by the transmutation of Jesus the prophet and teacher into the divine Christ, the only Son of God.

When Christianity first burst into life, Jesus was acclaimed by his followers to have been the long-awaited Jewish Messiah (*Christos*, in Greek) but under the direction of Paul, a Hellenistic Jew, he was raised to divine status. So what some religious historians would categorise as a personality cult evolved within two or three centuries into the widespread Christian religion. The homely and down-to-earth teaching of Jesus about the coming Kingdom of God metamorphosed into a divinely conceived cosmic plan for the salvation of humankind. In Christian imagination Jesus had ascended to the heavenly places to sit at the right hand of God the Father, ready for the Last Judgment that, among other things, would result in the destruction of the present world and its replacement with 'a new heaven and a new earth'. This scenario had largely replaced the teaching of Jesus within two or three centuries, and it has remained the core of Christian orthodoxy for nearly two millennia.

But, as the Jewish sages said, nothing lasts forever. Indeed, at the very time when Christian orthodoxy was being set in stone as never before with the construction of the mighty cathedrals that still dominate European cities, the seeds of the future humanism preserved within Christianity were nourished by the injection of a new cultural element that came by way of Islam. As we noted in the last chapter, it was during the Middle Ages that the influence of the Muslim Averroës and the rediscovery of Aristotle's philosophy of nature turned Western thought onto the new path known as the *via moderna*. It was from Islam that some in the Christian West began to explore the physical sciences and to learn with wonder about the teachings of Aristotle, Ptolemy and Pythagoras. The simplest example of this new direction is the replacement of the clumsy Roman numerals with what we still call the Arabic numeral system, an essential change for the subsequent development of European science.

The newly obtained knowledge of Aristotle encouraged the Franciscan Friar Roger Bacon (c. 1214–92) to study the physical world in a way that was independent of revealed theology. Keen to enquire into "the ways of God", he became an experimentalist, assembled a primitive telescope and invented the thermometer. He believed that by observing the succession of events in nature we could propose a general law to account for them and then proceed to test the validity of that law. He called this procedure a universal experimental principle, and through his writings the term *experimental science* became widespread in the West. He strove to create a universal wisdom that embraced all the sciences and was organised by theologians. Out of his deep Christian conviction he came to believe that a better understanding of the natural world would serve to confirm the truth of the Christian religion. He rebuked his fellow priests for their ignorance and amassed a great deal of information on all kinds of subjects, including alchemy. Yet Roger Bacon was an erratic genius who could be incredibly naïve, and by later standards his work left much to be desired.

Unlike Bacon, Nicholas Copernicus (1473–1543) and Galileo Galilei (1564–1642) were much more like modern scientists. Copernicus was a Polish priest and scholar who, on the basis of mathematical calculations and astronomical observations, arrived

at a theory of the solar system that challenged and revolutionised the biblical understanding of the universe. His theory was
confirmed by Galileo and built upon by Johannes Kepler (1571–
1630) and Isaac Newton (1642–1727). In the same period
Francis Bacon (1561–1626) laid the philosophical foundations
for empirical science in his *Advancement of Learning* and has
justifiably been called its father. In respect to all of these pioneers
of the scientific approach, it was their strong belief in one divine
Creator which underpinned their basic conviction that natural
phenomena operated in a rational and comprehensible way.

The reason the emergence of empirical science did so much to
promote humanism is that it gave budding scientists confidence
in their ability to add to the growing pool of knowledge without
depending on past tradition and divine revelation. Thus, the growing scientific method was providing humans with a new, humanly
based way by which to acquire both reliable knowledge and new
understanding. Before long, as the case of Galileo exemplifies, this
new methodology challenged and at last undermined traditional
knowledge and ancient perceptions of the cosmos. In particular it
challenged the authority of the Bible, though not until the nineteenth century did the Bible become a fallen idol.

Though initiated by priests, the practice of science soon became a lay pursuit. Even so, most early Western scientists were
devoutly Christian. Isaac Newton wrote more about religious
issues than about science. The Royal Society, founded for improvement of natural knowledge, had many clergymen in its
ranks of amateur scientists, but as the authority wielded by empirical science increased, the authority attached to the priesthood
began to wane—and never more rapidly than in the twentieth
century.

Along with the rise of empirical science, another important
cultural movement likewise promoted humanism and became
known as the Renaissance (a name it received only in the mid-
nineteenth century) because it was shaped by the rebirth of

interest in the pre-Christian classical literature of the Greeks and Romans. Indeed, even though the period after the fall of Rome was the very time during which Christianity spread throughout Europe, it came to be known as the Dark Ages, a label coined by Francesco Petrarch (1304–74), the Italian poet and philosopher subsequently heralded as 'the Father of Humanism'. The term *humanist* that was adopted by the leaders of the Renaissance did not mean for them what it means today, but it was a step in that direction. While remaining a devout Catholic, Petrarch saw no conflict between religious faith and the nurturing of the human potential to think rationally, as the Greek and Roman men of letters had so clearly done. By encouraging the study of the Greek and Roman classics, he helped to promote the intellectual flowering that characterised the Renaissance.

Desiderius Erasmus (1466–1536), said to be the most learned man of his time and labelled 'the Prince of the Humanists', was a Dutch priest and theologian who exerted a powerful influence throughout Europe through his writings. Partly because he shrank from open dissension and partly because he thought adequate reform of the church could be managed only from within, he refused to join Martin Luther and the other Protestant Reformers even though he was highly critical of the late mediaeval church. Instead of openly challenging traditional Christian doctrine, he therefore chose to subject it to playful satire as he did in his little book *In Praise of Folly*, where he held up to ridicule some of the traditions and superstitions of the Catholic Church. For example, he wrote of monks in the monastery that "They bray like donkeys in church, repeating by rote psalms that they have not understood, imagining they are charming the ears of their heavenly audience with infinite delight". This became his best-known book, even though he had scribbled it down rather quickly while staying a few days with his friend Sir Thomas More. No wonder the Catholic Church came to condemn him as the one who "laid the egg that Luther hatched". His little book illustrates how trivial and unintended events have sometimes had the widest effect.

At about the same time a learned and somewhat precocious 24-year-old Italian philosopher wrote *Oration on the Dignity of Man*, a work often called the Manifesto of the Renaissance. Pico della Mirandola (1463–94) contended that human beings were the last creatures to be created so that they could come to know the laws of the universe and admire its greatness. His essay clearly expressed the growing self-confidence that marked humanism.

With the recent invention of the printing press, the ideas of all these voices criticising the status quo were soon able to spread more widely and rapidly than ever before, and thus the speed of cultural change began to accelerate.

Hard on the heels of the Renaissance came the dissenting voices that promoted the Protestant Reformation. This development fragmented the unity of Western Christendom and fostered the growth of *nationalism*—the latter offering a secular way of giving personal allegiance to a higher authority. For Protestants the focus of ultimate authority was transferred from the Papacy to the Bible. But the Bible is open to different interpretations, and this, together with the rise of nationalism, explains why there never arose a single international Protestant church parallel to the Roman Catholic Church. The Renaissance and the Reformation both opened the door to and encouraged more independent thinking.

The emerging independent thinking came to a head in the eighteenth-century Enlightenment, a cultural movement in which human rationality finally prevailed over the supposed divine authority channelled through the Pope (for Catholics) or the Bible (for Protestants). The philosopher Immanuel Kant (1724–1804) described the Enlightenment (also called the Age of Reason) as humankind's final escape from the bondage that resulted from obligatory acceptance of the thoughts of others, whether in the form of secular traditions or religious pronouncements. In the Enlightenment humans had at last become free to think for themselves and, for better or for worse, to appeal to human reason as the final arbiter.

Although the appeal to reason has not proved to be the sure and swift avenue to truth that was promised, the freedom to think produced astonishing results in the post-Enlightenment period for it opened the door to the modern period of accelerating cultural change. It marked the end of the Age of God and our victorious entry into the humanistic age. As faith in a caring and providential deity declined, the cultural vacuum it left behind was filled by an increasing faith in mankind's ability to solve all the problems it encountered.

An important example of this developing humanistic ethos was the recognition of the human origins of the Bible. Following the publication of David Friedrich Strauss's *The Life of Jesus* (1835), freedom from the dominant idea of divine revelation enabled scholars to study the Bible as a library of humanly composed documents that reflected both the beliefs and the prejudices of their writers' times. By the beginning of the twentieth century a flood of new biblical commentaries was appearing, nearly all of them based on the new modes of study known as textual, historical and literary criticism. What had long been viewed as divinely revealed truth was now seen to be of human origin. No longer subservient to the words of Holy Scripture, scholars gained new mastery of it by understanding how it originated and why only an interactive interpretation could make it relevant to a much later age.

By the mid-nineteenth century the fruits of critical study of the Bible, together with Darwin's scientific theory of biological evolution, led to fierce debate and bitter divisions in the Christian world. This in turn gave rise to a widespread belief (still common today) that science and religion are in fundamental conflict. This unfortunate impression was spread by two books: *History of the Conflict between Religion and Science* by John William Draper

(1874) and Andrew Dickson White's two volumes on *History of the Warfare of Science with Theology in Christendom* (1896).

This supposed conflict should not be read back into the early days of science, when the exact opposite was true. The first scientists were themselves priests, for by and large it was only by becoming a priest that a university education might be accessed. Even more important, it was the monotheistic basis of Christianity that supplied the emerging physical sciences with the axioms essential to their enterprise: If God created everything, and did so for a purpose, there must be a consistency underlying and permeating all natural phenomena, however diverse they may appear to be. Because of the monotheism expressed so clearly and succinctly in the opening chapter of the Bible, everything was thought to flow from one creative source and constitute a *uni*verse. The pioneering efforts of Roger Bacon made it the task of the physical sciences to discover the very nature and workings of the universe that constitute the 'ways of God'.

This is why some of today's scientists (such as Einstein and Paul Davies) rather surprisingly refer to God; they are using the word symbolically and not with its traditional connotation. The reason for this is that, as theologian Gordon Kaufman pointed out, in our thinking the concept of God has long served as a unifying point to which everything else can be oriented. And so Davies entitled his book on "science and the search for ultimate meaning" *The Mind of God*.

It is no accident, then, that the modern scientific way emerged out of a monotheistic culture. In his Gifford lectures on *The Relevance of Science* in 1959, the German physicist and philosopher Carl Friedrich von Weizsäcker pointed out that the concept of a single divine creator had supplied the essential basis for the emergence of modern science. Thus, the beginnings of empirical science in the Middle Ages not only mark the first steps in the transition from the monotheistic age to the humanistic age, but illustrate how and why the one evolved out of the other and show the essential connection between the two; the concept of one Creator God had come to serve as a symbol affirming the unity of the universe. No wonder Weizsäcker deplored the way in which "the church was blind to the true nature of modern times and the modern world was equally blind to its own nature. The modern world was the result of the secularisation of Christianity".

The concept of evolution began to surface in the Western mind just when people were becoming more deeply aware of how everything changes over the course of time. This was partly because cultural change was accelerating so quickly that it was now clearly noticeable within the span of a human lifetime. And surprising as it may seem, only at this late date did historians recognise the need to define and study what they at first called Modern History—that is, *all* history after the ancient Greeks and Romans! Geologists also made people aware of an ever-changing landscape, and finally, the publication of Darwin's *On the Origin of Species* added a vast new dimension to the growing sense of history. By the end of the nineteenth century most highly educated Westerners had become aware that they lived in an exceedingly old and ever-changing world. But this brave new world was not to be feared, they thought, for their fast-developing technology gave them great confidence that humanity would increasingly gain control over nature. By the year 1900 it seemed quite clear that the humanistic age had arrived and was to be welcomed.

Perhaps the first person to announce the coming of the new humanistic age, and certainly the one who most clearly understood its significance and relation to the Christian past, was the nineteenth-century theologian and philosopher Ludwig Feuerbach (1804–72). Even his most severe critic, the noted theologian Karl Barth, said of him in 1925, "No philosopher of his time penetrated the contemporary theological situation as effectually as he did, and few spoke with such pertinence." When only thirty-seven years of age, Feuerbach published his epoch-defining *The Essence of Christianity* (1841), and soon followed it with *The Philosophy of the Future* (1843) and *The Essence of Religion* (1851).

Feuerbach's place in the evolution of human religion may be compared with that of Copernicus in cosmology and that of Darwin in biology. As Copernicus revolutionised our understanding of the universe and Darwin revolutionised our understanding of our origins, Feuerbach revolutionised our understanding of religion when he quite literally turned the world of religious thought upside down or, as he said, "the right way up". In the monotheistic age, religious thought started with God and moved to humankind; Feuerbach started with humankind and moved to God. As he said, "The old world made spirit the parent of matter, the new makes matter the parent of spirit."

It is quite remarkable that Feuerbach reached this insight nearly twenty years before Darwin's idea of evolution became widely known. We may say that Feuerbach simply took Occam's philosophy of nominalism to its logical conclusion. He recognised that religious concepts like 'God' were humanly created ideas with humanly created names. Whereas the old world said that God had made humankind, the truth is that humans have made 'God'.

But Feuerbach did not disparage the idea of God, as the atheists did and still do. He went beyond the debate between theist and atheist and reached a much more positive conclusion by recognising the real essence of the word God and the role it had played. Feuerbach claimed that the reality of 'God' consists of the attributes assigned to the imagined deity, as illustrated by the Bible's assertion that "God is love". The attributes that comprise 'God' represent the projection of our highest moral qualities, such as love, justice, compassion and forgiveness onto a cosmic backdrop. Even when the imagined divine 'being' disappears, these values remain. Though they are human values, it is to them that we must respond as our forbears responded to God.

Further, he said, we also project onto 'God' all the abilities which we humans would like to possess, such as power, knowledge, ubiquity, durability, etc. There they become the divine attributes of omnipotence, omniscience, omnipresence and eternity. I believe this explains why the most fervent believers in God are also the ones who are most certain that they know exactly what God is thinking and planning, for it is what they inwardly think and aspire to.

Feuerbach contended that as soon as God is recognised to be a humanly constructed idea, it becomes meaningless to ask whether God exists or not. For that reason Feuerbach refused to call himself an atheist. He contended that the real question, the one that most deserves a thoughtful answer, is: How did the concept of God evolve in human thought and what role or purpose has it served in cultural history? This question is exactly what the previous chapter attempted to answer.

Unlike the average atheist of the past or present, Feuerbach believed that both 'God' and religion, when properly understood, play an important part in human self-understanding. The role of God is to gather under one symbolic term all the moral values to which we feel bound to respond, along with all the laws of nature to which we are bound to submit. All of these in their sundry particular ways 'lord it over us' or are as 'God' to us. Together they explain the human condition. In Feuerbach's view, *theology* (the study of God) is really *anthropology* (the study of humankind). It is the study of our highest human values and of how we can make the most of our lives.

But Feuerbach's crowning *tour de force* is found in the way he focused on the central Christian doctrine—the incarnation—and made it the key that unlocked the door to the humanistic age. This doctrine asserts that God came down to Earth and became incarnate in Jesus of Nazareth. Its true significance, argued Feuerbach, is that 'God' had indeed become enfleshed in the human condition in a way exemplified by Jesus of Nazareth, but not exclusively by him. Christianity was good news because it opened the way for humankind to live life to the full by seeking to embody its highest values. For too long humankind had bewailed its sinfulness and impotence when in truth the problem was our having become divorced from human values by projecting them onto a mythical being in the heavens. The victorious Christ figure, seen as God incarnate in human flesh, was a symbol of the return of those values to their human source, thus leaving the mythical divine throne empty. That is why he titled his book *The Essence of Christianity*.

Feuerbach began his book by affirming the importance of religion, the practice of which he saw as the chief characteristic distinguishing humans from all other animals. In his later Heidelberg lectures he called on his audience to embrace what he called the religion of man, saying:

> We must replace the love of God by the love of man as the only true religion, . . . the belief in God by the belief in man, i.e., that the fate of mankind depends not on a being outside it and above it but on mankind itself. . . . My wish is to transform friends of God into friends of man, believers into thinkers, devotees of prayer into devotees of work, candidates for the hereafter into students of this world, Christians who, by their own profession and admission are half-animal, half-angel, into men, into whole men.

Feuerbach was so far ahead of his time that he was not only dismissed from the university and excommunicated by the church, but was so widely and bitterly condemned in the public arena that he was soon forgotten. Yet now at the beginning of the twenty-first century many of his observations are so commonplace that we fail to realise how outrageous they seemed when he first made them. To be sure, Feuerbach was overly optimistic about what life would be like in a world where we humans can no longer appeal to divine help as our forbears believed they could. In the humanistic age we have entered we must accept the fact that we have no one else to turn to but one another in seeking answers to our questions.

I began this book by posing the question: Where did we come from? The mind-boggling story of the evolving universe has supplied the answer. Even though this story began far in the past, so far in fact that our minds have great difficulty in coping with the time span involved, it is only in very recent times that we have been able to tell it. This has become possible thanks to the combined efforts of many different classes of scientists—chiefly cosmologists, geologists, biologists, anthropologists and archaeologists. It cannot be overemphasised that it is only by pooling the skills, knowledge and wisdom of many different individuals, past and present, that we can find satisfying answers to the most basic questions of human existence.

We humans are on our own as we find ourselves living in an uncaring universe, one which is evolving by chance and necessity. Like it or not, this is the human predicament into which the evolutionary process has placed us. Let us now turn to the issues of how we are coping with our predicament, starting with the question of who we are.

# The Human
Situation

# CHAPTER 10

# Who Are We?

*This book started by raising the question of where we came from.*

The story of evolution that has just been sketched is our current answer and it applies equally to all of us, no matter how diverse our personal or racial characteristics may make us appear. Moreover, this story supersedes all the many traditional and scriptural stories of our origin and forms the basis for any emerging global culture of the future. True, this story bears a certain resemblance to the creation story that appears in the first chapter of Genesis, which I have occasionally alluded to. However, this resemblance is not wholly surprising when we remember that the modern science that produced the evolutionary story was born within the cultural context of the Christian West.

The self-evolving cosmos is our common story of origins. It is the new Great Story through which to understand life. We humans have come to be the creatures we are as a result of the energy that burst forth from the Big Bang 13.75 billion years ago. Out of that singularly explosive event there gradually formed the myriad galaxies of stars that still continue to evolve within the ever-expanding continuum of space and time. Within one galaxy (which we often call "ours") a supernova explosion about 4.5 billion years ago led to the formation of our solar system of eight planets revolving round one newly born star (*our* sun). Of those eight it was only on planet Earth that simple forms of life began to emerge from about 3.5 billion years ago. For some two billion years they were so microscopically small as to be quite invisible to the human eye—had we been there to observe them. But these simple, numerous forms of earthly life eventually developed in complexity until eventually they became the myriad plant and animal species that came to inhabit Earth. Most of those are now extinct and known only by their fossil remains, yet the many forms of life that remain still amaze us with their diversity. They include a large family of mammals that diverged from their common ancestors sixty-six million years ago, and from them emerged the primates out of which came the genus *Homo*. Even this genus diversified, but *Homo sapiens* is the only species to have survived. Thereafter, this species spread around the globe, slowly creating the more than six thousand languages and cultures that tend to disguise our common ancestry. We humans have arrived at this point in cosmic time by a very long and tortuous route, punctuated by chance events of catastrophic magnitude.

This story of where we came from helps us to understand who we are. We are not spiritual beings imprisoned in physical bodies, creatures who, as Plato thought long ago, are temporarily exiled from our true home. We are physical creatures made from the substance of the Earth. The ancient biblical myth had it about right when it declared that we are made from dust and to dust we will return. We are earthlings, the products of planet Earth, who have evolved through innumerable genetic changes from the earliest and simplest forms of life. We are genetically connected with all other forms of earthly life from the chimpanzee right back to the single-celled bacterium. Our bodies still carry many vestiges of the prolonged biological journey that our genes have taken through the successive stages of fish, amphibian, reptile, mammal and hominid. Indeed, neurologists who trace the evolution of the human brain speak of its most primitive part as 'the reptilian brain'. Our bodies have been shaped by the physical conditions of Earth, and we remain dependent on the Earth for our continued existence. Even if we attempt to venture out into space we have to take with us such components as air, water and food in order to survive.

Then why is it that until 150 years ago, when this story of the universe began to be told, we humans believed ourselves so different from all other animals as to be separated by an unbridgeable gulf? Indeed, it was long believed in the Christian world that we humans were of such a high order of beings that we were made in the image of God, the creative source of everything. The reason for this notion of 'special creation' is to be found in our culture, for not only are we the product of biological evolution but we also reflect the eons of cultural evolution that followed. Compared to both cosmic and biological evolution, cultural evolution has been remarkably rapid. We measure the life of the cosmos in billions of years and that of animal life in millions, but the journey of our species from caveman (whose pattern of life was not unlike that of today's gorillas and chimpanzees) to that of today's civilised human beings has taken less than one hundred thousand years.

This rapid advance was made possible by the human invention of language. Thus, humans are to a unique degree a self-made species. This may seem extremely hubristic, but far from basking in the thought, we must consider the great responsibilities it places on our shoulders.

Further, not only do our physical bodies mark us as Earthlings, but our minds show us to be historical creatures. Just as the genes we inherit shape our bodies, the thought worlds of the cultures into which we are born shape our minds—almost totally early in life, and very considerably even to the end of our days. Moreover, since each culture's history is constantly unfolding, each of us reflects cultural elements that were dominant at the time of our birth: we belong forever to the century in which we live, and only a very few, by virtue of their insights, may be said to have been born before their time.

But however great our current sophistication relative to all other creatures, we humans remain animals and, like all the rest, each of us experiences a single, finite existence between conception and death. To be sure, our genes possess a sort of immortality that we do not, for they live on in the regeneration process. Our species not only persists from generation to generation but may conceivably evolve even further. Still, each of us is allotted but one life, which occupies only a tiny fragment of cosmic time. We do not survive death in any form at all, except in the memories of those who knew us. This follows from the recognition that we are psychosomatic creatures and the inflexible corollary that no form of personal life can follow the death of the body— a notion long contemplated by many religious traditions of the past.

One of the values of pondering the story of evolution is a deeper understanding of what sort of creatures we humans are. On the one hand, we occupy so infinitesimally small a place in the cosmos that its sheer immensity in both space and time is more than we can grasp; the world we are accustomed to is what we can see with the naked eye. Yet as insignificant as we earthly creatures are, our high level of consciousness and our rapidly expanding thought world allow us a limited mental image of the universe and the ability to ruminate on all that has evolved within the universe since its beginning.

Such is the creative power of this self-evolving universe that it has not only given us life and enabled us to think (as it has all the other animals) but it has also enabled us to remember, to know and most of all to plan purposefully. This leads to a breathtaking thought, one that should fill us with wonder and awe: though we are merely the products of the evolving cosmos and made of the dust of Earth, we can see and begin to understand the universe from which we are made. We can almost imagine the impersonal universe looking at itself through our eyes and attempting to understand itself through our minds.

An ancient psalmist who stumbled on this realisation more than two thousand years ago, being a theist and a man of his time, expressed it this way:

> When I look up to the heavens,
> the work of your fingers —
> the moon and the stars
> that you set in their place,
> I ask, "What is humankind?
> "Why should you remember it?
> "Why should you care for mortal humans?"
> For you have made them little less than gods.
> With glory and honour you have crowned them.
> (Psalm 8, my translation)

This feeling of awe, once experienced by our spiritual forbears with respect to the supposed Creator God in the heavens, has not vanished with the 'death of God'; it has simply been transferred to the self-organising cosmos itself, for that is what has brought us forth. We see the same universe they did, but through a very different cultural lens. Ancient talk of the gods and later talk of God reflected the way our ancestors personalised the forces of nature by projecting their own subjective attributes onto the world around them. What most amazes us about the self-organising universe is perhaps its propensity to bring forth ever-more complex wholes. As discussed in Chapter 3, Jan Smuts called this creative power of the universe *holism* and explained that this universe is not a static and changeless thing, but rather a self-creative and evolving process. The process of evolution is itself the supreme miracle—by which I mean it is a phenomenon that leaves us in wonderment.

A corollary and nearly equal miracle is that we humans are part of this process and have evolved to become the thinking, feeling, planning and questioning creatures that we are. And if it seems to have taken the universe a very long time to bring us forth, that is largely because our arrival was not the result of any original design or purpose on the part of the universe. We are here largely by chance!

This is something we may at first find difficult to accept, for it certainly undermines any feeling of self-importance, yet calm reflection soon makes it clear that, as individuals, we are certainly here by chance. For however much people in love like to believe they were meant for each other from the beginning, we need think only of the many chance events that led up to the meeting and mating of our parents to realise that it is more by accident than by design that each of us came to be born. As we come to understand the biology of human conception it becomes clear that the DNA that determines our individual physiology and personality results from a chance meeting of an egg from our mother with one of thousands of possible sperm from our father. And this is but one example at a personal level of the innumerable chance events that have taken place on the cosmic scale. It was certainly no purposeful event that brought an end to the Age of the Dinosaurs, and that was only one of the more dramatic occurrences in the long story of geogenesis and biogenesis. Yet if that particular global catastrophe had not left an environmental niche for mammals to flourish and diversify in, we humans would not have evolved and spread over the Earth as we have.

It is only because we *are* here and have collectively kept adding to our world of thought that we are now able to tell the story of evolution, a story that is far from complete for two reasons. First, new discoveries are constantly improving our knowledge of the past, and from time to time some dramatic new discovery changes our understanding of the cosmos or of the planet Earth, and thus of the origin and nature of life. Second, not only does the evolutionary process continue with no end in sight, but the evolution of human culture has now begun to accelerate at such a rate that its changes often leave us breathless.

Having been derived from knowledge available to us humans and told from a human point of view, this story of evolution inevitably contains a distinctly human bias. Nor can it be otherwise in a cosmos that contains no absolute reference point in either time or space. All that we think and say remains relative to the conditions in effect where and when we attempt to put our story together.

As we look back in time from our present vantage point on this planet and observe what seems to be the universe's strong propensity to form itself into ever-more complex units—and eventually to bring forth the human being, the most complex entity of all—we are tempted to interpret this holistic proclivity as evidence of a mastermind at work. It long appeared to provide indisputable evidence for the existence of God; philosophers called it the argument from design.

In the latter part of the twentieth century the promoters of what is called Creation Science referred to the source of order in the universe as Intelligent Design. The trouble is that their system is not science at all, but rather an ideology posing as science; indeed, it starts from the conclusions it seeks to defend (mainly derived from biblical affirmations) instead of starting with the observable data and allowing them to lead to a logical conclusion. Their theological arguments from design also fail to acknowledge that examples of design (which certainly exist in abundance) are the result of evolution by natural selection; biological evolution proceeds on the basis of what best promotes the survival of the species. Of the innumerable chance mutations that take place in the regenerative process, only those that prove beneficial to the species become permanent features. Thus, survival value is what has led to more complex and—to us—more wonderful designs. Clearly, this phenomenon does not demand the presence of a designer; rather, the design is a by-product of this self-organising universe.

Scientists have discovered that the cosmos is so finely tuned that if certain basic physical factors—for instance, the gravitational constant, the mass of the proton and the age of the universe—had been even slightly different, the evolutionary process could not possibly have produced the human species. The observation that the conditions on this planet happen to be just right for our emergence has been dubbed the 'anthropic principle'. Some proponents of this principle claim that since it required no fewer than thirteen billion years to bring forth the human species, we have 'arrived', as it were, right on schedule! The anthropic principle (now divided into strong and weak versions) continues to be debated among both scientists and philosophers, but as yet it has failed to demonstrate that evolution proceeds according to any ultimate purpose. No doubt we would not be here if any of the basic constants had varied, but that hardly argues for an evolutionary process pre-programmed to produce us. A similar assertion would claim that a person wins a lottery because the balls were purposefully jumbled so as to fall in their favour. To conclude that the evolving universe exists for the purpose of producing us is simply another example of human hubris. Yes, the universe has produced us on planet Earth, but elsewhere in the universe there may be forms of life much wiser and nobler than we.

The Big Bang had the potential to produce a variety of universes, and the one that did evolve could and did produce us, but not for any preordained purpose. On the contrary, we emerged almost against the odds (recall the many extinct species of *Homo*). Any concealed purpose that we may imagine within the evolutionary process is no doubt the projection of our human fascination with the notion of purpose. As Nietzsche observed, just as the notions of good and evil originated in the human thought world, so did the idea of purpose. The evolutionary process is not inherently good or evil; neither does it pursue any purposeful path.

Whatever happens in cosmic evolution, from earthquakes to 'nature, red in tooth and claw' (as Tennyson says in his poem *In Memoriam AHH*, 1850), occurs beyond and irrespective of human moral judgments of good and evil. When we judge some cosmic events to be good and some to be bad we are simply projecting our human value system—something we have created in the course of cultural evolution—onto the cosmos. No eternal and absolute values waiting to be discovered by us exist in the universe.

We humans created the notions of good and evil and all non-human creatures are quite unaware of them. We may judge the tiger cruel for catching, killing and eating its prey, but the animal is simply ensuring its survival. Of all the earthly species, only humans live in a world where our notions of good and evil apply. The myth of origins found in Genesis 2 quaintly acknowledges this by depicting a time when our primeval ancestors first encountered the notions of good and evil. We read that after Adam and Eve had eaten of the forbidden fruit, the gods exclaimed, "Mankind has become like one of us, knowing the difference between good and evil". Moral sensitivity was regarded as a divine attribute! It is highly ironic that theologians long interpreted this mythical event as 'the Fall', when the myth itself portrayed it as the arrival of a higher level of awareness. In the context of the new evolutionary story of our origins it should, as Georg Hegel (1770–1831) was among the first to observe, be recognised not as the fall of mankind but as its rise.

We judge to be good whatever enhances *our* well-being, including *our* very existence; and we judge to be evil whatever injures *our* well-being. This is why the natural phenomenon of death was for so long treated as an enemy that had to be overcome, whereas now we acknowledge death to be essential to the processes of life. Indeed, death is just as important as birth and has made the evolution of species possible. So that we might be here today, countless generations before us had to die, after passing on their genes, occasionally mutated. Once again it is clear that our value system has evolved within our ever-developing thought world and is constantly changing and evolving. But having developed and applied it to ourselves, we tend to project it onto other forms of life and even onto the cosmos itself.

The way we humans project our notions of purpose and morality is neatly illustrated in the biblical myth of creation (Genesis 1), which states that "God saw everything that he had made, and behold it was very good." The very concept of God as a cosmic creator had itself evolved from our ancient forbears' hopes of finding purpose and meaning in the universe; then, in attributing the above declaration to God, they were projecting their own value judgments into his mind.

That the evolutionary process does not reflect some prior cosmic purpose was recognised long ago by the Greek philosopher Democritus (c. 460–370 BCE) when he said, "Everything existing in the universe is the fruit of chance and of necessity"—"necessity" being equivalent to what we today call the laws of nature. This dictum was given renewed expression by the Nobel Prize-winning geneticist Jacques Monod in his *Chance and Necessity* (1971). Monod had even greater reason to emphasise the importance of chance, for biologists have shown that chance mutations in the regenerative process are the key to evolution and the origin of species.

Strangely enough, and perhaps as a consequence of the recent phenomenon in Western culture referred to as the 'death of God', the role of chance is being increasingly accepted by the popular mind and is reflected in our colloquial language. We have long spoken of events as 'happenings', a word that clearly implies chance, but in recent times the word *luck* has been appearing more frequently in our daily speech. When a chance event is advantageous for us we call it good luck; when disadvantageous, it is called bad luck. And whereas our forbears used to bid each other farewell with "God be with you" or "God bless" (the implication being that we would be cared for by an invisible guiding hand), we now commonly wish each other "Good luck" or "Have a nice day", by which we offer the hope that events will prove advantageous to their well-being—while simultaneously acknowledging that events could very well turn out otherwise.

Thus, the path followed by evolution through some thirteen billion years has been directed solely by a combination of chance and necessity—until the emergence of the human species! Now for the first time, planned human activity has given the process a limited yet ever-expanding degree of purposeful direction. And this injection of purpose is greatly accelerating the evolutionary process everywhere on the planet.

This new factor in the cosmic equation arose with the emergence of consciousness, which introduced the possibility of choice. For a very long time choice was a virtually negligible element, and it still remains of minimal importance among non-human species. Most animal activity is directed by instincts that have evolved because of their survival value. Even so, while a bird's instinct directs it in the construction of its nest, the element of consciousness in its tiny brain is responsible for a vast number of choices in selecting specific twigs for the general task.

As the complexity of brain and nervous system increased, the level of consciousness attained became higher and the more significant the role of choice became. As noted earlier, human consciousness has reached a level greatly superior to that of our nearest extant relatives and language has enabled us to create a thought world that could be transmitted to the next generation. This continually evolving thought-world then nurtured the planned and purposeful behaviour that began to show itself increasingly on Earth.

Early on, for example, our ancestors mastered the use of fire to serve their purposes—first for cooking and keeping warm, later for protection and clearing land, and at last for refining ores and shaping metals. They also invented tools of a complexity far

exceeding the simple products of other species in order to ease
their labour and magnify its results. Largely by the medium of
language-based culture, humans have advanced step-by-step from
primitive stone tools to today's complex technology. In doing so,
we have not only introduced purposeful creativity into the evolutionary process, but have greatly accelerated the process itself.

Before we became aware of the story of our origins, we drew
a clear distinction between human constructions and those of
insects, birds and animals. We judged it to be natural for a bird
to build a nest but artificial (or unnatural) for humans to construct bridges and cathedrals. Relative to the universe, however,
these activities are of the same order. It is as natural for humans
to construct skyscrapers as it is for birds to build nests. Much of
the difference between them lies in the complexity of the final
product, but even more important, perhaps, is the component
of consciously designed purpose. For conscious creatures are not
simply products of the evolutionary process; indeed, their very
consciousness (especially in the case of humans) introduces purposeful direction into a process that previously proceeded solely
by chance and necessity.

After the mastery of fire and the invention of tools, the next
significant example of human purposefulness appeared in the
cultural transition made by our ancestors a little over ten thousand years ago, when many of them decided (made the choice!)
to give up being hunters and gatherers in order to become agriculturalists. Abandoning a nomadic existence that was wholly
dependent on the food offered by nature, they took up a settled
existence, increasing their supply of food by assisting the natural
processes of growth and minimising such elements of chance as
the vagaries of the climate. By planting seeds, choosing the most
useful grains and selectively breeding the most fruitful plants
and animals, they harnessed the natural processes of growth to
produce more and better food both for themselves and their
domestic animals. Nearly all the fruits and grains we use today

have resulted from the myriad improvements to natural growth consequent on human planning.

A settled life led to permanent dwellings, and nomadic tribal life (necessarily limited in numbers) gave way to villages and city-states, an expanding population, division of labour and specialised occupations. But with the development of new skills and interests came an increasing degree of mutual dependence. The personal ties of kinship that held tribal societies together were gradually replaced by the mutual responsibilities and loyalties that hold the nation together, and we still regard our nationality as an attribute of our identity.

Between 1750 and 1850 there was a giant cultural leap forward that is known as the Industrial Revolution. The application of steam power (to be quickly followed by electric power, the internal-combustion engine and nuclear power) greatly increased both the quantity and the diversity of what humans could manufacture. This second great revolution began in England and quickly spread to Western Europe, North America, India and Japan; soon most of the world was caught up in what proved to be the first step in the now rapidly advancing process of globalisation. In addition to affecting nearly every aspect of life, including communication and travel, it changed human society from being chiefly rural and agricultural to being predominantly urban and disconnected from its food sources. Increasing dependence on 'factory farms' and manufactured goods led to a population boom and unprecedented growth in wealth and international trade. All these factors propelled the advent of the modern global world by making industrialised nations increasingly dependent upon one another for the health of their economies.

The interests and activities of the human race are no longer predominantly devoted to the production of food to ensure its survival. Although non-human creatures spend much or all of their time seeking food and caring for their young, most humans now enjoy an ever-increasing amount of leisure time that they use to pursue arts, games and entertainment. Unfortunately, the material and cultural benefits bestowed by the modern world are far from being equally shared, for there is an ever-increasing gulf between the rich and the poor, not only within nations but between what we now call First World and Third World nations. What follows here applies chiefly to First World nations.

Today people live much more complex and sophisticated lives than they did only a hundred years ago. Due in part to a knowledge explosion it has been necessary to provide formal education for all, increasingly up to the tertiary level. The diversification of the sciences, a rapid advance in technology, and a multiplicity of new occupations and other human pursuits all mean that we live in a rich and rapidly expanding thought world. We inhabit electrified homes; we listen to recorded music; we watch films; we are in instant contact with one another on phone and computer. Further, we regularly engage in air travel, both domestically and around the world. All this has emerged during the last century, but only in the last thirty years have the computer and the Internet revolutionised the way we do our work and communicate with one another. The accumulated knowledge of humankind has become available to everyone through the click of a computer mouse.

Perhaps the most significant advances have been in the area of human health. Life expectancy has approximately doubled in the last two hundred years as a result of reduced infant mortality, improvements in personal hygiene, breathtaking advances in medical science and the virtual elimination of a number of major diseases by widespread vaccination programs that have aided and improved our already well-developed natural immune systems.

But medical science now does much more than simply nurture and promote the natural healing process: At an accelerating rate it is developing new practices and products for the preservation and improvement of human life. Not only do we benefit from such assistive technology as prostheses, dentures, implant lenses and hearing aids, but failing body organs—kidneys, hearts, lungs and livers—can be transplanted from living donors or even cadavers! The arcane process of conception can now be conducted in laboratories; we can exercise a degree of control over the process of death. Issues of life and death were long considered to be the exclusive prerogative of God, but it would appear that the human species is now developing skills that come close to matching those of the supposed creator of the universe.

Moreover, since humankind has now managed to visit the moon and is sending out probes into the furthest reaches of the solar system, who can guess what the human species may yet achieve both here and in outer space? At least where this planet is concerned, we humans constitute the leading edge of the evolutionary process. For just as the evolving universe produced human consciousness and humans created the concept of purpose, so the element of purpose is being increasingly introduced into the hitherto-blind process of evolution and is greatly accelerating the process.

We now know where we humans come from and we have a much clearer idea of who we are. Though we have evolved from bacteria, we have become like the gods our primitive ancestors created in their imaginations. It is quite understandable why the biblical writers thought we were made in the image of God. Indeed, it is conceivable that, as Christian myth would have it with the man Jesus in the first century, in the twentieth century it is the human race itself that is becoming the incarnation of the God created by human imagination, for humans are now achieving on Earth what was once thought to be solely the work of God in heaven.

From 1900 onwards the people of the Western nations looked into the future with abundant confidence. With the new tools of science and technology, what would they not be able to accomplish? The sky was the limit and science fiction pioneers like

H G Wells (*The War of the Worlds*) and Arthur C Clarke (*2001: A Space Odyssey*) described the future exploits of humankind in space. To be sure, the twentieth century saw many advances—some in areas not even imagined in 1900—but relatively few of those imagined by science fiction.

Indeed, by 2000 the mood had noticeably altered, for in the course of only one hundred years the prospects of the human future had changed dramatically. Though human civilisation has reached what must seem its zenith, and some populations enjoy a previously unimagined quality of life, it was becoming clear that all is not well with life on Earth. Two world wars, the nuclear bomb, brutal acts of ethnic cleansing from the Holocaust onward, the threat of an all-out nuclear war and finally the threat of global warming and catastrophic climate change all place the future of humanity in serious doubt.

By 2000 the Western world no longer dominated the globe; its empires had disintegrated. The Christian missionary attempt to evangelise the world had ground to a halt, and was in rapid retreat in many areas. The Christian world had long drawn its inspiration from scriptures that included a divine injunction to "Be fruitful and multiply and fill the earth and subdue it and have dominion over the fish of the sea and over the birds of the air and over every living thing that moves upon the earth"—that commandment has been quite literally fulfilled, but with dire results. Our species has overfilled Earth and hardly knows how to stop increasing. Before 1900 the human population stood in symbiotic harmony with other species and with the environment. The impact of human activity on Earth's surface was negligible and, if for some reason human habitation of an area ceased, nature soon took over, often obliterating the signs of prior human life.

Apart from the new steam-driven factories and railways, early twentieth-century people were still living in the days of the horse and cart, the barrow and shovel and simple plough, as their ancestors had for ten thousand years. But a century later all that had changed. Not only had we quadrupled in number, but we had developed mechanised technology that enabled us to change the contours of Earth and master the waterways. We filled the air with radio waves, discovered how to use nuclear energy and invented electronic tools at an accelerating rate. We now dominate the planet in a way no other species has ever done. We treat planet Earth and its creatures as if they are ours. We take possession of the land and exploit its resources for our own use. We mass-produce some animals for our food while carelessly causing the extinction of others. Worst of all, we are constantly polluting our air and water, the two most important commodities on which our survival depends.

In short, we now find to our dismay that we are continually at war, not only with one another but with the planet from which we have emerged and on which we depend. Planet Earth, our only home, is far from being the paradise hoped for in 1900, and the road ahead is far from clear. An increasing number of thoughtful observers speak of dire crises to be faced in the imminent future. While still glorying in our human achievements, we are now forced to speak of the human predicament. Where are we going? To this final question we now turn.

# Where Are
# We Going?

*Since every story has both a beginning and an end, what
will be the end of the new Great Story?*

That, of course, is something we cannot know, particularly now
that the components of consciousness, choice and subsequent
alternatives have entered the evolutionary process through the
emergence of the human species. Nevertheless, once a story is
well on its way, we often know enough of the characters and the
unfolding plot to have some idea of where it may be heading.
On this basis we can attempt to look into the future and imagine
the possible destinies of the human race. We individuals, being
mortal, have no destiny beyond that of our finite existence here
on Earth, but it is within our power to undertake an exploration
of the future of the human species and of the cosmos as a whole.

Let us start with the larger picture. It seems ironic that we
can be more confident in sketching the future of the cosmos
over the next few million years than we can be in forecasting
what will be the headlines in tomorrow's newspapers. Indeed, as
physicists have found, we have even less success in predicting the
movement of subatomic particles at the quantum level. It seems
the larger the scale, the more the forces at work will continue
on their way without deviation. Our present knowledge of the
cosmos encourages us to believe that it will continue to expand
in the foreseeable future much as it has been doing since it came
into being with the Big Bang. But for how long?

That is a question that scientists are discussing with great interest. The celebrated cosmologist Lawrence Krauss humorously

confesses in his book *A Universe from Nothing* that he took up the study of cosmology because he "wanted to be the first person to know how the universe would end". From the Big Bang onwards, the evolving universe has been shaped by two primary, opposite forces: the momentum given to cosmic matter by the explosive nature of the Big Bang and the force of gravity pulling it all together again (which Teilhard referred to these as tangential energy and radial energy, respectively).

We observe the operation of these opposing forces whenever we launch a rocket into space. If the force propelling the rocket is not strong enough to overcome Earth's gravity, then the rocket falls back to Earth. If the force is stronger than gravity, then it proceeds out into space, travelling on its tangential path forever unless it comes under the influence of another object in space. If the rocket's propelling force is exactly equal to that of gravity, then it becomes a satellite which continuously orbits Earth.

As the rocket has three possible trajectories, likewise on the grand scale the expanding universe has three possible futures. Cosmologists refer to these futures as open, closed or flat. In a closed universe, gravity will eventually halt the current expansion and then initiate the reverse process, causing the universe to collapse in on itself and eventually end in what has been called the Big Crunch. An open universe, on the other hand, is one that will keep on expanding forever. A flat universe is one whose expansion continually slows down but never quite comes to a halt. The current view of Krauss and some other cosmologists is that we inhabit a flat universe. Interesting though this question is to the enquiring scientific mind, it is not one that would seem to have any bearing on the future of the human species.

This is not the case, however, when we turn to the future of planet Earth. Gravity will ensure it continues to orbit the sun as one of its satellites for a very long time to come. But as the sun uses up the fuel in its nuclear furnace, it will slowly expand to become a red giant so that in some four billion years it will swallow up the nearest planets. While that prospect is so far away as to cause us no alarm, we do need to be alert to another danger: Earth is being continually bombarded with objects from outer space.

It has been calculated that asteroids as large as ten kilometres across hit Earth every fifty to one hundred million years. The last of these was the asteroid impact in Mexico about sixty-six million years ago, an event that is now considered responsible for the demise of the dinosaurs. Smaller asteroids are more frequent but do less damage. The Barringer crater in Arizona was made about fifty thousand years ago by an asteroid some one hundred metres in diameter. While it may be extremely unlikely, we can never rule out the possibility of Earth being struck at any time by an asteroid large enough to destroy human civilisation and probably bring an end to the human species altogether.

It was this very possibility that supplied the theme for a mid-twentieth-century science fiction film *When Worlds Collide*. It is interesting to note that the film's introduction made reference to the biblical story of the Great Flood, in which God was prepared to blot out humankind from the face of the Earth. This reminded us that even the biblical tradition contemplated the possibility that the human species might someday cease to exist. Theists, believing that every cosmic event occurs for a moral reason, are able to claim that the Great Flood was sent by God as punishment for human wickedness.

Nevertheless even the biblical legend portrayed God as one who rued the drastic nature of his decision and so gave Noah careful instructions on how to build an ark in order to save both his family and breeding pairs of all animals and thus make a new start. *When Worlds Collide*, on the other hand, adopts the modern humanistic world view in which divine help could not be expected to prevent a perfectly natural event; this left humans solely dependent on their own resources, so they set to work designing and building a spacecraft in the very limited time before the expected impact. Only a few could be saved from the upcoming catastrophe, and these were to be chosen by ballot at the last moment, a situation that gave rise to much drama. As I remember the film, all too little thought was given to the question of where they were going and whether they would find a planet with living conditions comparable to those of the planet they were abandoning.

While both biblical legend and film drama belong to their respective genres of myth and science fiction, they may be taken to symbolise a very real and imminent crisis that the human species now faces. This one, however, will not be caused by an 'act of God' or a chance collision with an object from outer space. A number of futurologists believe that we are in danger of bringing about our own demise simply by continuing our present patterns of behaviour.

The possibility of such a fate first came to my attention when I was researching for my book *The World to Come* (1999). As we approached the new millennium it was becoming clear that the dominance of Christianity had come to an end and that we were entering what could be called the Global Era. I even made the rash suggestion that the year 2001 CE could be appropriately relabeled 1 GE, as the first year of the Global Era. After all, the current Western calendar goes back to the now-dubious calculations of the Christian monk Dionysius in the sixth century, and has never been universally accepted even in the Christian world, let alone the world at large. Today we could well do with a common world calendar that would allow all people to acknowledge their common humanity and increasingly global civilisation.

What will the new Global Era bring? I discovered a number of problems of which I had previously been only vaguely aware which are now accelerating in intensity and have the potential to lead us towards a climax of catastrophic dimensions as early as the twenty-first century. With these in mind I sketched ten possible scenarios for the future of the human race in the twenty-first century. The rise of global terrorism was number eight in my catalogue, and we had hardly moved into new millennium when the now infamous 9/11 of 2001 brought to our TV screens the destruction of the Twin Towers in New York. We are reminded of the continuing terrorist threat every time we board a plane and have our luggage and even our persons searched.

Another scenario I sketched is the rise and spread of pandemics. These have long been the scourge of the human race: the Black Death of the fourteenth century killed about quarter of the population of Europe, and the Spanish flu of 1918–20 killed more people than the Great War that had just ended. Then came AIDS and the avian flu. In spite of great advances in medical science, pandemics remain high on the list of major threats to the human species because viruses can mutate more rapidly than appropriate vaccines can be produced. As international travel has greatly increased, new viruses can spread around the globe much more quickly than before. Even more threatening is the possibility that knowledge of this minute form of life might end up in the hands of those who would make new and deadly viruses to be used as weapons. Whereas all pre-2000 epidemics were caused by natural pathogens, biotechnology can now create new agents of deadly mass infection.

Terrorism and pandemics are but two dangers that threaten the well-being of the human race in the Global Era. In 1992 Paul Elkins, a research economist in London University, discerned four domains of human activity that presented such grave risks to the human future that he described them as Four Holocausts: war and militarisation, denial of basic human rights by oppressive governments, economic destitution in the form of mass poverty among a fifth of the world's population, and environmental destruction that will gradually render the planet uninhabitable.

I suspect he may have regarded them as the modern equivalents of the Four Horsemen of the biblical Apocalypse. He chose the biblical term Holocaust because it had become associated with the most shocking human atrocity of modern times—the extermination of six million people in Nazi death camps. No doubt Elkins wished to shock us into awareness of the threats he perceived. Another researcher, Derek J Wilson, adopted Elkin's meaning for the term holocaust for the title of his 500,000-word volume Five Holocausts (2001), in which he gathered a massive amount of information which supported his idea of five holocausts. He saw his fifth—the population explosion—as the underlying cause of all the others.

The first of Elkins' four holocausts was militarism, including nuclear warfare. Indeed it was during the Cold War between Russia and the Western world in the 1960s that we first imagined the possibility of bringing about an apocalyptic end to the human race by waging an all-out nuclear war. In 1982 Jonathan Schell discussed the dire consequences of such a war in his book *The Fate of the Earth*. The possibility of a nuclear holocaust by no means disappeared with the end of the Cold War, as we see in the international uneasiness resulting from so many countries having nuclear weapons. Nuclear weapons continue to proliferate in spite of desperate efforts to restrict them, and we see the increasing danger of them falling into the hands of terrorists. Their advent was a Pandora's Box that can never be closed, and the threat they pose is a sword of Damocles forever hanging over our heads.

A quite different concern comes from climatologists, geologists and other scientific experts: that we humans have been endangering our own future simply by ignoring the consequences of what we have been doing to the Earth. The celebrated historian Arnold Toynbee devoted his last years to the writing of *Mankind and Mother Earth*. Published in 1976, it ends with these chilling words: "Will mankind murder Mother Earth or will he redeem her? . . . That is the enigmatic question that now confronts Man."

Earth has long been appropriately referred to as "Mother", for not only has it brought forth all the life we know, but we continue to depend upon it for our existence. We need the air, water and food it provides; without them, we die. Each species has evolved in a symbiotic relationship with its environment and its survival depends on how well it adapts to the ever-changing conditions of the planet. Failure to do so quickly leads to extinction.

We humans have had a basic awareness of this ever since we evolved consciousness. Long before we formulated the law of gravity we had learned to respect it, knowing full well that if we jumped off a cliff we were likely to be killed. And when our ancient forbears developed agriculture, they learned to attune their practices to the changing seasons and the conditions of the soil. They knew little of the complex nature of the Earth's ecology, but that mattered little so long as the consequences of human activity on the surface of the Earth remained relatively inconsequential in comparison with the forces of nature.

That is no longer so, for two main reasons. First, in the last 150 years the sudden massive increase in human population has placed a great burden on the Earth's capacity to supply us with food and other essentials for life. Second, the greatly increased power afforded by modern technology has allowed us to alter the natural conditions of Earth that are conducive to our survival, often without realising it. These two facts alone mean that human habitation on this planet is entering an entirely new era in its evolutionary history, one in which our actions now endanger the future of all living species, including our own.

The results of our maltreatment of Earth are all the more serious because at first they did not set off the kind of alarm bells that the testing of nuclear weapons did. The dangers have been creeping up on us quietly and we are only gradually becoming aware of them—and then more because of what the scientists tell us than because of what we can readily observe.

One of the first warning signs came in the 1930s, when we learned that modern farming methods were exacerbating soil erosion, causing a yearly loss of thousands of tonnes of soil that rivers carried to the sea. Then in 1962 Rachel Carson alerted us to the damage that insecticides had been doing to the environment; her book Silent Spring sparked the environmental movement in America.

Soon after that we became aware of a growing variety of problems on several fronts: the loss of arable land due to desertification, the depletion of oceanic fish stocks from overfishing, river pollution and a diminishing supply of fresh water. Perhaps worst of all, we have been polluting the atmosphere with carbon dioxide and methane gases, causing global warming and potentially disruptive climate change.

It is mainly during the last decade that the public has become increasingly aware of the phenomenon of global warming and its alarming consequences. Former American Vice President Al Gore did much to better inform us though his film An Inconvenient Truth. Accelerating climate change is causing the seas to warm and melt masses of ice in Greenland and the polar caps. The resulting rise in sea levels will eventually inundate large portions of such nations as Holland and Bangladesh and coastal cities everywhere. While these effects may not become widely apparent until the end of the century, we have already experienced an increase in the intensity of both droughts and storms. It seems likely that the Earth will soon be unable to feed the growing population of a planet where millions are already underfed.

Although international conferences continue to be held in response to the problem of climate change, very little has so far been achieved. Even the well-informed governments of affluent countries cite economic reasons for their reluctance to take the drastic measures necessary even to retard the onset of global warming. It may well require a global catastrophe to impel us to undertake the radical changes necessary to avert global disaster.

Moreover, whereas primitive human cultures evolved slowly in a symbiotic relationship with nature, the theistic culture from which industrialisation emerged has long directed our attention beyond Earth to another world, all too often teaching us to see ours as a fallen world. Whether Jewish, Christian or Muslim, theists have long accepted without question that humanity's chief obligation was to obey the commandments of its divine Creator. While much in those faith traditions remains highly commendable, teaching followers to respect and love one another, they are seriously deficient regarding respect for Earth and nature. Indeed, they have often had the effect of degrading Earth and encouraging us to treat it as ours to exploit and dominate.

Yet another concern is our highly complex national and global economies. Before the Industrial Revolution there was little need to mention 'the economy', for most people subsisted on what they produced on their small patch of earth. Today most of us live in cities, and because we depend on the easy and efficient interchange of products, often from the other side of the world, our very lives depend on the incredibly complex network of goods, services and useful employment that we call the economy. Our politicians and economists tell us that this system requires continuous growth in order to remain healthy and serve us well.

Yet in 1972 a report entitled *Limits to Growth* issued by the global think tank, the Club of Rome, declared that because all natural resources such as oil, gas and water are in finite supply, it is not possible for economic growth to continue indefinitely. But that means that the economic goals and policies pursued by nearly every nation are on a collision course with the limitations imposed by a finite earth. Even apart from this economic contradiction, the interplay of national economies within the global economy appears to be encountering problems inherent to the capitalist system. In 1998 international financier George Soros argued in his book *The Crisis of Global Capitalism* that we were already in the early stages of a global bear market that would inevitably lead to a global recession, a worldwide depression and finally the disintegration of the capitalist system. William Greider observed in his book *One World, Ready or Not* that capitalism inevitably causes the rich to get richer and the poor to get poorer. This gross imbalance in wealth becomes more obvious every year, not only within nations but also between the developed nations and the developing nations—so much so that we now refer to them as First World and Third World countries. In his book *Earth Currents* (1995) Howard Snyder forecast a global crisis around the year 2020. Even as I write, the National Intelligence Council of the United States has just issued their report on Global Trends, which concludes that "the world is at a critical juncture in human history" and declares a number of countries to be at high risk of economic collapse before 2030.

There is therefore little doubt that because of these factors— nuclear weapons, global warming, uncertainty in the global economy, the spread of terrorism and the likely rise of new pandemics—the Global Age finds the welfare of the human race under threat on a scale never before experienced.

Being global problems, these can be solved only by the joint efforts of the entire human race. In the past we humans lived in our own independent worlds, geographically isolated from one another and culturally diverse. Today we are being drawn together into one world whether we like it or not. Languages, cultures and religious beliefs that have hitherto evolved in isolation are now intermingling and competing with one another as former lines of demarcation disappear. Greatly improved communication, increased trade, migration and tourism have all brought humans closer together. We are increasingly moving into one world, hence the appropriate term globalisation. Not surprisingly, this has already resulted in cultural, political and military conflicts around the globe, and even two world wars. We are now caught up in a global maelstrom of unprecedented and accelerating cultural and economic change.

Of course, globalisation can bring about good as well as bad; it can lead to unbelievable catastrophes but it can give birth to a richer worldwide culture by cross-fertilising ethnic and national cultures and producing an ever-expanding body of knowledge. But to achieve these ends we must first solve the grave and complex problems outlined above. Unfortunately we have yet to forge the commonality of vision, purpose and effort essential to achieving a common good. Humankind now consists of over seven billion people, most of whom are chiefly concerned with their own personal, local or national interests. Those who are already underfed—estimated to number upwards of a billion—are understandably primarily concerned with the immediate problem of where their next meal is coming from. And the majority of the world's people remain dangerously unaware of the global dangers lurking ahead.

For these and similar reasons some futurologists have issued dire warnings about the human future. In 2003 Martin Rees, President of the British Royal Society and Astronomer Royal, went so far as to give the ominous title of *Our Final Century* to a book contending that humanity has at best a fifty-fifty chance of surviving to the end of the twenty-first century.

How ironic it is that just when we are beginning to pay more attention to scientists and scholars than to the priests and theologians, they should be warning us that humankind's future is under threat! Indeed, it is doubly ironic, for it was the threat of a catastrophic future that was midwife at the birth of Christianity. In the first century a confluence of factors led some to conclude that the present universe was about to end and that God would replace it with "a new heaven and a new earth". These people strongly believed that they lived in the 'last days' (eschata in Greek, a word that gave rise to the term eschatology). Biblical scholars now speak of the eschatological expectation that pervades the New Testament writings. Fear of what would happen to them at what they imagined to be the imminent end of the world gave birth to the cry "What must I do to be saved?" Christianity originated in part as the answer to that question, and as it spread and developed within the context of Graeco-Roman culture, the initial question widened into that of how to be saved from divine judgment.

In this respect it is interesting to observe the extent to which the language of salvation (along with its near synonym "conservation") has come back into common use in recent decades. It started with such slogans as Save the whales, Save the dolphins, Save the black robin; then came the more general Save endangered species. Climatologists have even used the catchcry, Save the Earth! Whereas in the first century people sought salvation from the end of the world through from God's judgment, today we rightly fear the consequences of our hostility towards others and the damage we have been inflicting on the environment that sustains us.

The words of Toynbee quoted earlier have become increasingly timely. In spite of our advanced technology—indeed, partly because of it—we humans are in danger of bringing about our own demise. The words that the biblical author put into the mouth of Moses have become strangely relevant once again: "I have set before you this day life and good, death and evil. . . . Therefore choose life." Of course these words were not addressed to individuals, they were addressed to communities. They were addressed to a community, in this case the people of Israel who hoped to continue on indefinitely from generation to generation. It was the lives of their grandchildren and later descendants that would be most affected by the choices they were challenged to make.

And so it is with us. As the acronym MAD (Mutually Assured Destruction) makes clear, nuclear war would endanger the future of the human race. Similarly, if we fail to take the drastic measures necessary to halt global warming, it is not we but our descendants who will suffer. The injunction to love one another has taken on a whole new dimension, for it now includes not only our contemporaries but the billions yet to be born.

We may well conclude that the responsibilities now resting on our shoulders exceed both our understanding and our readiness to acknowledge them. This is why Toynbee declared that only the rise of a new and appropriate religion could generate the mass willpower needed to reverse the human destruction of Earth and begin to build a viable and sustainable future. Is such a religion possible? What would it be like? Even the term "religion" is suspect in this humanistic age because of its long association with a now-outmoded supernatural world.

Since the modern secular world emerged out of the Christian West, Christianity may well make a significant contribution to the new religion. Some Christian spokespeople have already expounded what they call Secular Christianity, and Don Cupitt (as mentioned in Chapter 9) referred to the teachings of Jesus as the New Religion of Life. In any case, if we think of religion as a compelling conviction (which is probably what Toynbee had in mind) then we need go no further than today's conservation enthusiasts, devotees of saving endangered species, Green political parties and similar idealistic visionaries to see the beginnings of the naturalistic religion of the future. Unlike religions of the past, it has no founder, yet it shows certain parallels with them. Where adherents of past religions felt obliged to obey laws revealed by a divine source, today's devotees feel compelled to respond positively and obediently to what the natural world demands of us. In other words, they believe that we should cease dominating the Earth, squandering its riches and polluting the air and water, and instead we must turn our attention to treasuring the Earth, wisely husbanding its resources, caring for its inhabitants and learning how to live in a symbiotic relationship with the planet.

The basis for this embryonic religion of the future is the new Great Story that I have tried to tell in this book. We must not only allow it to replace the biblical story of origins and its counterparts in other traditions, but also use it to inspire us with the sense of awe and holiness that permeated the former religions. The telling and retelling of the Great Story can motivate us to take appropriate action as we learn more of the grandeur of the cosmos and of the miraculous complexities and designs observable in the biosphere. As I have so often said above, the evolutionary process is a truly awe-inspiring phenomenon.

One of the first to be gripped by the Great Story and to interpret it religiously was Teilhard de Chardin. He synthesised science and religion into a total breathtaking vision. Moreover, he saw the already-current globalisation (he called it *planetisation*) as a turning point in human evolution and one that was rich with possibilities. While always acknowledging that evolution has proceeded by chance and necessity and not because of any inherent guiding force, he nevertheless discerned in the story what he called a "cosmic drift" towards an unknown goal he labelled *Omega*.

Evidence of this drift can be found in the ongoing tendency of the universe to form ever-more complex wholes, referred to previously as holism. Let us briefly recapitulate the scenario: Soon after the Big Bang, the universe was formed from unstable subatomic particles the stable wholes we call atoms. Then atoms united in the new and more complex wholes that we call molecules, and molecules became ever-more complex, eventually becoming mega-molecules that at last united to form the first living cells. All this took eons, just like the later process in which simple cells united to form organisms and organisms became increasingly complex until they evolved into animals.

But the holistic tendency did not stop there. Since most forms of life beyond the cellular stage were of two genders, the mating of the two sexes, followed by the nurturing of the young, introduced the beginnings of the yet more complex whole we call the family. Among humans, families coalesced into clans, tribes and nations—societies held together by bonds of kinship and common loyalties. Even such insect societies as hives of bees and nests of ants work so closely together for the common good that we can imagine that the individuals are continually guided by a unifying spirit.

In this cosmic tendency to form ever-more complex wholes, Teilhard observed two opposing movements in operation, which he called divergence and convergence (similar to tangential energy and radial energy). When mutations in DNA cause a genus to divide into two or more species, we have an example of divergence. When the mutating gene becomes so widely shared within the evolving group as to produce what is called the 'pure-bred' of the species, we have an example of convergence.

Similarly, when Homo sapiens survived the extinction of all the other species of Homo, it spread around Earth, displaying divergence by diversifying into numerous racial and ethnic groups. Now that our species has spread over the Earth in abundance and has nowhere else to go, it is showing convergence as its previously independent groups intermingle. Thus globalisation, caused by the very finiteness of Earth, marks a significant turning point in human evolution.

In globalisation Teilhard saw the human species beginning to fold back upon itself, merging all ethnic groups and cultures into one unified human species and thus creating one new, global culture. The human species, he surmised, was on the threshold of becoming a super-society within which the collectivity of consciousnesses would form a higher level of consciousness that he dared to call super-consciousness. As the individual human consciousness develops a centre we call the ego or self, so there would arise in the collective super-consciousness "a spiritual centre, a supreme pole of consciousness, upon which all the separate consciousnesses of the world may converge".

No doubt many will feel that Teilhard has led us into a fantasy land, a realm of science fiction, but his thoughts are well worth pondering, and I include an outline of them in this concluding chapter because few creative thinkers have been able to tell the Great Story—and from it create a vision of the future—as consistently as he did. In any case, when we attempt to enquire into future possibilities, we do well to rein in our critical faculties. Who, for example, would have thought in 1900 that a century later we would be flying around the world, viewing people on the other side of the world on television screens, sending men to the moon and transplanting body organs? Who would have thought we would be linking ourselves electronically through the Internet, Facebook and Twitter—even finding life partners through these 'social media'? Is this an indication of our becoming a 'collectivity of consciousnesses'? And what else may lie ahead?

Moreover, at a time when gloomy but realistic voices are warning us that the human species may well be bringing about its own demise by one or more of the crises discussed above, Teilhard's thoughts can encourage us to find in the evolutionary process an unexpected reason for hope. He fully conceded that, since cosmic evolution has produced the human species by chance and necessity rather than from any purposeful intention, there is no guarantee that a glorious future will eventuate. In spite of this, the clear evidence of cosmic holism suggested to him that the best Earth can produce may be yet to come.

Of course if we regard the evolutionary process as a source of hope and put our ultimate faith in it (as Teilhard certainly did) it would seem tantamount to bringing back something like the discarded God of monotheism. It is true that though he remained a devout Catholic to the end of his days, Teilhard virtually came to think of the evolutionary process as God. He went so far as to say that if he lost his former faith in God and in Christ he would still have faith in the Earth. To be sure, we need to be very circumspect when referring to the evolutionary process as God, if only because the traditional understanding of 'God' is still widespread. But if we recall that the humanly created concept of

God evolved as a way of unifying the unseen powers believed to transcend humanity, then it does make some sense. Indeed, that is just how process theologians like John Cobb were already interpreting the word God in the middle of the twentieth century. We are utterly dependent on the evolutionary process, and we are permeated by it. It is as God to us even though it is impersonal and therefore quite unlike the God of Abraham.

Moreover, as theologian Gordon Kaufman has pointed out, even in a secular world the term God can still perform a useful function in continuing to serve as "an ultimate point of reference" as it has done in the past. The important difference is that in recent times our ultimate point of reference is no longer the ancient sky-god but the Earth that brought us forth and the evolutionary process that brought forth the Earth. Our lives and our future as a human species are wholly dependent on this amazing, self-evolving cosmos. This led Kaufman to say, in his book *In Face of Mystery*, "To believe in God is to commit oneself to a particular way of ordering one's life and action. It is to devote oneself to working towards a fully humane world within the ecological restraints here on planet Earth, while standing in piety and awe before the profound mysteries of existence." That may well come to be regarded as a summary of the religion of the future.

In sum, the future of the human race remains an open question. On the one hand we must take full account of the perilous crises already facing us; like black clouds on the horizon, they indicate an imminent period of storms that could lead to catastrophic outcomes. It does seem unlikely that humans worldwide will be able to muster the willpower and the unity of action to avoid them altogether. On the other hand, we can draw hope from the Great Story of how we came to be here at all. It is a truly awe-inspiring universe that has brought us forth and, at least on this planet, has come to consciousness in us, displaying the human inventiveness, creativity and entrepreneurial skills that have helped to make us the creatures we are. And this potential may lead us to as-yet-unimaginable heights. If our descendants survive and evolve to reach an even more exalted state of being than ours, they will have arrived at what our forbears long aimed for when in their traditions (Buddhist, Jewish, Muslim or Christian) they hoped, respectively, to enter Nirvana, the Promised Land, the unity of all nations, or the Kingdom of God.

# Bibliography

Armstrong, Karen. *A History of God: The 4,000-Year Quest of Judaism, Christianity, and Islam*. London: Heinemann, 1993.

———. *The Great Transformation: The World in the Time of the Buddha, Socrates, Confucius and Jeremiah*. London: Atlantic, 2006.

Bacon, Francis. *The Advancement of Learning*. 1605.

Barrow, John D & Frank J Tipler. *The Anthropic Cosmological Principle*. Oxford: Oxford University Press, 1986.

Bellah, Robert. *Religion in Human Evolution*. Cambridge, Massachusetts: Harvard University Press, 2011.

Carson, Rachel. *Silent Spring*. Boston: Houghton Mifflin, 1962.

Confucius. *The Sayings of Confucius*. Translated by James R Ware. Toronto: Mentor, 1955.

Corballis, Michael C. *From Hand to Mouth*. Princeton: Princeton University Press, 2002.

———. *The Recursive Mind*. Princeton: Princeton University Press, 2011.

Cupitt, Don. *Taking Leave of God*. London: SCM Press, 1980.

———. *Jesus and Philosophy*. London: SCM Press, 2009.

Darwin, Charles. *On the Origin of Species*. London: John Murray, 1859.

Davies, Paul. *God and the New Physics*. London: Penguin Books, 1984.

———. *The Cosmic Blueprint*. Unwin Paperbacks, 1989.

———. *The Mind of God: Science and the Search for Ultimate Meaning*. London: Penguin Books, 1992.

Dawkins, Richard. *The Ancestor's Tale*. Boston: Houghton Mifflin Company, 2004.

Draper, John William. *History of the Conflict between Religion and Science*. New York: D Appleton, 1874.

Feuerbach, Ludwig. *The Essence of Christianity*. 1841. Translated by George Eliot (New York: Harper Torchbook, 1957).

———. *The Philosophy of the Future*. 1843. Translated by Manfred H Vogel (Indianapolis: Bobbs-Merrill, 1966).

———. *Lectures on the Essence of Religion*. 1851. Translated by Ralph Mannheim (New York: Harper & Row, 1967).

Frankfort, H & H A. *The Intellectual Adventure of Ancient Man*. Chicago: University of Chicago Press, 1946.

Geering, Lloyd. *Faith's New Age*. London: Collins, 1980. Revised and published as *Christian Faith at the Crossroads* (Santa Rosa, California: Polebridge Press, 2001).

———. *The World to Come*. Santa Rosa, California: Polebridge Press, 1999.

———. *Tomorrow's God*. Santa Rosa, California: Polebridge Press, 2000.

———. *Coming Back to Earth*. Salem, Oregon: Polebridge Press, 2009.

———. *Such is Life!* Salem, Oregon: Polebridge Press, 2010 and Wellington, NZ: Steele Roberts Publishers, 2010.

Gregory, Bruce. *Inventing Reality*. New York: John Wiley & Sons, 1988.

Greider, William. *One World, Ready or Not: The Manic Logic of Global Capitalism*. New York: Simon & Schuster, 1997.

Hawking, Stephen W. *A Brief History of Time: From the Big Bang to Black Holes*. New York: Bantam Books, 1988.

Hawking, Stephen & Leonard Mlodinow. *A Briefer History of Time*. New York: Bantam Books, 2005.

———. *The Grand Design*. New York: Bantam Books, 2010.

Huntington, Samuel P. *The Clash of Civilizations and the Remaking of World Order*. New York: Simon & Schuster, 1996.

Jeans, James. *The Mysterious Universe*. Cambridge: Cambridge University Press, 1930.

Kaufman, Gordon D. *In Face of Mystery: A Constructive Theology*. Cambridge, MA: Harvard University Press, 1993.

Krauss, Lawrence. *A Universe from Nothing: Why There Is Something Rather than Nothing*. New York: Free Press, 2012.

Leakey, Richard & Roger Lewin. *Origins Reconsidered: In Search of What Makes Us Human*. Boston: Little, Brown & Co, 1992.

Lovelock, James. *The Revenge of Gaia: Earth's Climate in Crisis and the Fate of Humanity.* London: Allen Lane, 2006.

———. *The Vanishing Face of Gaia: A Final Warning.* London: Allen Lane, 2009.

Lucretius. *De Rerum Natura.* 50 BCE. Translated by Sir Ronald Melville as *On the Nature of the Universe* (Clarendon Press, 1997).

Macquarrie, John. *In Search of Deity: An Essay in Dialectical Theism.* London: SCM Press, 1984.

Monod, Jacques. *Chance and Necessity: An Essay on the Natural Philosophy of Modern Biology.* London: Collins, 1971.

Popper, Karl with John Eccles. *The Self and Its Brain.* London; Boston: Routledge & Kegan Paul, 1977.

Rees, Martin. *Our Final Century: Will Civilisation Survive the Twenty-First Century?* London: Arrow Books, 2003.

Schell, Jonathan. *The Fate of the Earth.* New York: Knopf, 1982.

Smuts, Jan. *Holism and Evolution.* New York: MacMillan, 1926.

Snyder, Howard A. *Earth Currents: The Struggle for the World's Soul.* Nashville: Abingdon Press, 1995.

Soros, George. *The Crisis of Global Capitalism: Open Society Endangered.* New York: Public Affairs, 1998.

Stowe, Harriet Beecher. *Uncle Tom's Cabin.* Boston: John P Jewett, 1852.

Strauss, David Friedrich. *The Life of Jesus Critically Examined.* 1835. Translated by George Eliot (SCM Press, 1973 [1846]).

Swimme, Brian & Thomas Berry. *The Universe Story.* London: Penguin Books, 1994.

Teilhard de Chardin, Pierre. *Le Phenomenene Humain.* Paris: Seuil, 1955. Translated by Bernard Wall as *The Phenomenon of Man* (London: Collins, 1959).

Toynbee, Arnold. *Mankind and Mother Earth.* Oxford: Oxford University Press, 1976.

von Weizsäcker, Carl Friedrich. *The Relevance of Science.* London: Collins, 1964.

White, Andrew Dickson. *History of the Warfare of Science with Theology in Christendom.* London: MacMillan, 1896.

Wilson, Derek J. *Five Holocausts.* Wellington, NZ: Steele Roberts, 2001.

# Index

# About
# the Author

**Lloyd Geering** (D.D., University of Otago, New Zealand) is Emeritus Professor of Religious Studies at Victoria University of Wellington, New Zealand. A public figure of considerable renown in New Zealand, he is in constant demand as a lecturer and as a commentator on religion and related matters on both television and radio. He is the author of many books including *Such Is Life! A Close Encounter with Ecclesiastes* (2010), *Coming Back to Earth: From Gods, to God, to Gaia* (2009), and *Christianity without God* (2002).

In 2001, he was honored as Principal Companion of the New Zealand Order of Merit. In 2007, he received New Zealand's highest honor, the Order of New Zealand.

CPSIA information can be obtained at www.ICGtesting.com
Printed in the USA
LVOW07s0015210714

395003LV00002B/4/P

9 781598 151398